Micro State Management with React Hooks

Explore custom hooks libraries like Zustand, Jotai, and Valtio to manage global states

Daishi Kato

BIRMINGHAM—MUMBAI

Micro State Management with React Hooks

Copyright © 2022 Packt Publishing

Group Product Manager: Pavan Ramchandani

Publishing Product Manager: Aaron Tanna

Senior Editor: Hayden Edwards

Content Development Editor: Rashi Dubey

Technical Editor: Saurabh Kadave

Copy Editor: Safis Editing

Project Coordinator: Rashika Ba

Proofreader: Safis Editing

Indexer: Pratik Shirodkar

Production Designer: Vijay Kamble

Marketing Coordinator: Anamika Singh

First published: January 2022

Production reference: 2150722

Published by Packt Publishing Ltd.

Livery Place

35 Livery Street

Birmingham

B3 2PB, UK.

ISBN 978-1-80181-237-5

www.packt.com

For those who love coding.

– Daishi Kato

Contributors

About the author

Daishi Kato is a software engineer who is passionate about open source software. He has been a researcher of peer-to-peer networks and web technologies for decades. His interest is in engineering, and he has been working with start-ups for the last 5 years. He has been actively involved in open source software since the 90s, and his latest work focuses on developing various libraries with JavaScript and React.

I want to thank the people who have been close to me and supported me, especially my family, my co-workers, and contributors to my open source projects.

About the reviewer

Nalin Savara has worked on applications, games, web apps, and solutions since 1996. His work has spanned various platforms (PC/Mac, mobile, and web) and numerous technologies, including React.

He founded Darksun Tech and has contributed dozens of articles and book reviews to PCWorld India. He co-authored the book *Algorithm Analysis and Design*, which is used as a textbook by the Westwood Colleges (US) course on computer games programming.

Earlier in his career, Nalin was the lead architect of India's first 3D open-terrain game engine, and part of the team that developed Rightserve, Asia's first targeted banner advertising service. He also steered one of India's first cloud tutoring start-ups as its interim CTO.

Table of Contents

Part 2: Basic Approaches to the Global State

2

Using Local and Global States

3

Sharing Component State with Context

4

Sharing Module State with Subscription

5

Sharing Component State with Context and Subscription

Part 3: Library Implementations and Their Uses

6

Introducing Global State Libraries

7

Use Case Scenario 1 – Zustand

8

Use Case Scenario 2 – Jotai

9

Use Case Scenario 3 – Valtio

10

Use Case Scenario 4 – React Tracked

11

Similarities and Differences between Three Global State Libraries

Index

Other Books You May Enjoy

Preface

State management is one of the most complex concepts in React. Traditionally, developers have used monolithic state management solutions. Thanks to React Hooks, micro state management is something tuned for moving your application from a monolith to a microservice.

This book provides a hands-on approach to the implementation of micro state management that will have you up and running and productive in no time. You'll learn basic patterns for state management in React and understand how to overcome the challenges encountered when you need to make the state global. Later chapters will show you how slicing a state into pieces is the way to overcome limitations. Using hooks, you'll see how you can easily reuse logic and have several solutions for specific domains, such as form state and server cache state. Finally, you'll explore how to use libraries such as Zustand, Jotai, and Valtio to organize state and manage development efficiently.

By the end of this React book, you'll have learned how to choose the right global state management solution for your app requirement.

Who this book is for

If you're a React developer dealing with complex global state management solutions and want to learn how to choose the best alternative based on your requirements, this book is for you. Basic knowledge of JavaScript, React Hooks, and TypeScript is assumed.

What this book covers

Chapter 1, What Is Micro State Management with React Hooks?, explains how React Hooks help to deal with states. This allows us to have more purpose-specific solutions.

Chapter 2, Using Local and Global States, discusses two types of states. Local states are often used and preferable. Global states are used to share states between multiple components.

Chapter 3, Sharing Component State with Context, describes how Context is the primary method to deal with global states and how it works within React life cycles. We need some patterns to avoid extra re-renders.

Chapter 4, Sharing Module State with Subscription, explains how module state is another method for global state. It works outside React life cycles. We need to connect the module state to React components, but a Subscription to the module state makes it easier to optimize re-renders.

Chapter 5, Sharing Component State with Context and Subscription, shows another approach for the global state by using both Context and Subscription. It works within React life cycles and avoids extra re-renders.

Chapter 6, Introducing Global State Libraries, introduces some libraries with various approaches for solving common problems in global states.

Chapter 7, Use Case Scenario 1 – Zustand, discusses a library, Zustand, used to create a module state that can be used in React.

Chapter 8, Use Case Scenario 2 – Jotai, is about a library, Jotai, based on Context and the atomic data model. It can optimize re-renders too.

Chapter 9, Use Case Scenario 3 – Valtio, discusses a library, Valtio, for mutable module states. It automatically optimizes re-renders.

Chapter 10, Use Case Scenario 4 – React Tracked, discusses a library, React Tracked, used to enable the automatic render optimization for some other libraries, such as Context, Zustand, and React-Redux.

Chapter 11, Similarities and Differences between Three Global State Libraries, compares the three global state libraries – Zustand, Jotai, and Valtio.

To get the most out of this book

You will need a version of Node.js installed on your computer—v14 or later versions—and the `create-react-app` *package.*

Software/hardware covered in the book	Operating system requirements
Node 14	Windows, macOS, or Linux
React 17/`create-react-app` 4	Google Chrome
ECMAScript 2015/TypeScript 4	

Alternatively, an online code editor such as CodeSandbox can be used.

If you are using the digital version of this book, we advise you to type the code yourself or access the code from the book's GitHub repository (a link is available in the next section). Doing so will help you avoid any potential errors related to the copying and pasting of code.

It's highly recommended to create a small app based on what you learn in this book.

Download the example code files

You can download the example code files for this book from GitHub at `https://github.com/PacktPublishing/Micro-State-Management-with-React-Hooks`. If there's an update to the code, it will be updated in the GitHub repository.

We also have other code bundles from our rich catalog of books and videos available at `https://github.com/PacktPublishing/`. Check them out!

Conventions used

There are a number of text conventions used throughout this book.

`Code in text`: Indicates code words in text, database table names, folder names, filenames, file extensions, pathnames, dummy URLs, user input, and Twitter handles. Here is an example: "It would be nice to reuse the `Counter` component for different stores."

A block of code is set as follows:

```
const ThemeContext = createContext('light');

const Component = () => {
  const theme = useContext(ThemeContext);
  return <div>Theme: {theme}</div>
};
```

Bold: Indicates a new term, an important word, or words that you see onscreen. For instance, words in menus or dialog boxes appear in **bold**. Here is an example: "If you click the +1 button in **Using default store**, you will see that two counts in **Using default store** are updated together."

> **Tips or Important Notes**
> Appear like this.

Get in touch

Feedback from our readers is always welcome.

General feedback: If you have questions about any aspect of this book, email us at customercare@packtpub.com and mention the book title in the subject of your message.

Errata: Although we have taken every care to ensure the accuracy of our content, mistakes do happen. If you have found a mistake in this book, we would be grateful if you would report this to us. Please visit www.packtpub.com/support/errata and fill in the form.

Piracy: If you come across any illegal copies of our works in any form on the internet, we would be grateful if you would provide us with the location address or website name. Please contact us at copyright@packt.com with a link to the material.

If you are interested in becoming an author: If there is a topic that you have expertise in and you are interested in either writing or contributing to a book, please visit authors.packtpub.com.

Part 1: React Hooks and Micro State Management

In this part, we introduce the concept of micro-state management, which gained attention with React hooks. We also cover the technical aspects of the `useState` and `useReducer` hooks to get ready for the next part.

This part comprises the following chapter:

- *Chapter 1, What Is Micro State Management with React Hooks?*

1
What Is Micro State Management with React Hooks?

State management is one of the most important topics in developing React apps. Traditionally, state management in React was something monolithic, providing a general framework for state management, and with developers creating purpose-specific solutions within the framework.

The situation changed after **React hooks** landed. We now have primitive hooks for state management that are reusable and can be used as building blocks to create richer functionalities. This allows us to make state management lightweight or, in other words, micro. **Micro state management** is more purpose-oriented and used with specific coding patterns, whereas monolithic state management is more general.

In this book, we will explore various patterns of state management with React hooks. Our focus is on global states, in which multiple components can share a state. React hooks already provide good functionality for local states—that is, states within a single component or a small tree of components. **Global states** are a hard topic in React because React hooks are missing the capability to directly provide global states; it's instead left to the community and ecosystem to deal with them. We will also explore some existing libraries for micro state management, each of which has different purposes and patterns; in this book, we will discuss Zustand, Jotai, Valtio, and React Tracked.

> **Important Note**
> This book focuses on a global state and doesn't discuss "general" state
> management, which is a separate topic. One of the most popular state
> management libraries is Redux (`https://redux.js.org`), which uses
> a one-way data model for state management. Another popular library is XState
> (`https://xstate.js.org`), which is an implementation of statecharts,
> a visual representation of complex states. Both provide sophisticated methods
> to manage states, which are out of the scope of this book. On the other hand,
> such libraries also have a capability for a global state. For example, React Redux
> (`https://react-redux.js.org`) is a library to bind React and Redux
> for a global state, which is in the scope of this book. To keep the focus of the
> book only on a global state, we don't specifically discuss React Redux, which
> is tied to Redux.

In this chapter, we will define what micro state management is, discuss how React hooks allow micro state management, and why global states are challenging. We will also recap the basic usage of two hooks for state management and compare their similarity and differences.

In this chapter, we will cover the following topics:

- Understanding micro state management
- Working with hooks
- Exploring global states
- Working with `useState`
- Using `useReducer`
- Exploring the similarities and differences between `useState` and `useReducer`

Technical requirements

To run code snippets, you need a React environment—for example, Create React App (`https://create-react-app.dev`) or CodeSandbox (`https://codesandbox.io`).

You are expected to have basic knowledge of React and React hooks. More precisely, you should already be familiar with the official React documentation, which you can find here: `https://reactjs.org/docs/getting-started.html`.

We don't use class components and it's not necessary to learn them unless you need to learn some existing code with class components.

The code in this chapter is available on GitHub at `https://github.com/PacktPublishing/Micro-State-Management-with-React-Hooks/tree/main/chapter_01`.

Understanding micro state management

What is micro state management? There is no officially established definition yet; however, let's try defining one here.

> **Important Note**
> This definition may not reflect community standards in the future.

State, in React, is any data that represents the **user interface** (**UI**). States can change over time, and React takes care of components to render with the state.

Before we had React hooks, using monolithic state libraries was a popular pattern. A single state covers many purposes for better developer experience, but sometimes it was overkill because the monolithic state libraries can contain unused functionalities. With hooks, we have a new way to create states. This allows us to have different solutions for each specific purpose that you need. Here are some examples of this:

- Form state should be treated separately from a global state, which is not possible with a single-state solution.

- Server cache state has some unique characteristics, such as refetching, which is a different feature from other states.

- Navigation state has a special requirement that the original state resides on the browser end and, again, a single-state solution doesn't fit.

Fixing these issues is one of the goals of React hooks. The trend with React hooks is to handle various states with special solutions for them. There are many hook-based libraries to solve things such as form state, server cache state, and so on.

There's still a need for general state management, as we will need to deal with states that are not covered by purpose-oriented solutions. The proportion of work left for general state management varies on apps. For example, an app that mainly deals with server states would require only one or a few small global states. On the other hand, a rich graphical app would require many large global states compared to server states required in the app.

Hence, solutions for general state management should be lightweight, and developers can choose one based on their requirements. This is what we call micro state management. To define this concept, it's lightweight state management in React, where each solution has several different features, and developers can choose one from possible solutions depending on app requirements.

Micro state management can have several requirements, to fulfill developers' various needs. There are base state management requirements, to do things such as these:

- Read state
- Update state
- Render with state

But there may be additional requirements to do other things, such as these:

- Optimize re-renders
- Interact with other systems
- Async support
- Derived state
- Simple syntax; and so on

However, we don't need all features, and some of them may conflict. Hence, a micro state management solution cannot be a single solution either. There are multiple solutions for different requirements.

Another aspect to mention regarding micro state management and its library is its learning curve. Ease of learning is important for general state management too, but as the use cases covered by micro state management can be smaller, it should be easier to learn. An easier learning curve will result in a better developer experience and more productivity.

In this section, we discussed what micro state management is. Coming up, we will see an overview of some hooks that handle states.

Working with hooks

React hooks are essential for micro statement management. React hooks include some primitive hooks to implement state management solutions, such as the following:

- The useState hook is a basic function to create a local state. Thanks to React hooks' composability, we can create a custom hook that can add various features based on useState.

- The useReducer hook can create a local state too and is often used as a replacement for useState. We will revisit these hooks to learn about the similarities and differences between useState and useReducer later in this chapter.

- The useEffect hook allows us to run logic outside the React render process. It's especially important to develop a state management library for a global state because it allows us to implement features that work with the React component lifecycle.

The reason why React hooks are novel is that they allow you to extract logic out of UI components. For example, the following is a counter example of the simple usage of the useState hook:

```
const Component = () => {
  const [count, setCount] = useState(0);
  return (
    <div>
      {count}
      <button onClick={() => setCount((c) => c + 1)}>+1
      </button>
    </div>
  );
};
```

Now, let's see how we can extract logic. Using the same counter example, we will create a custom hook named useCount, as follows:

```
const useCount = () => {
  const [count, setCount] = useState(0);
  return [count, setCount];
};
```

```
const Component = () => {
  const [count, setCount] = useCount();
  return (
    <div>
      {count}
      <button onClick={() => setCount((c) => c + 1)}>
        +1
      </button>
    </div>
  );
};
```

It doesn't change a lot, and some of you may think this is overcomplicated. However, there are two points to note, as follows:

- We now have a clearer name—useCount.

- Component is independent of the implementation of useCount.

The first point is very important for programming in general. If we name the custom hook properly, the code is more readable. Instead of useCount, you could name it useScore, usePercentage, or usePrice. Even though they have the same implementations, if the name is different, we consider it a different hook. Naming things is very important.

The second point is also important when it comes to micro state management libraries. As useCount is extracted from Component, we can add functionality without breaking the component.

For example, we want to output a debug message on the console when the count is changed. To do so, we would execute the following code:

```
const useCount = () => {
  const [count, setCount] = useState(0);
  useEffect(() => {
    console.log('count is changed to', count);
  }, [count]);
  return [count, setCount];
};
```

By just changing useCount, we can add a feature of showing a debug message. We do not need to change the component. This is the benefit of extracting logic as custom hooks.

We could also add a new rule. Suppose we don't want to allow the count to change arbitrarily, but only by increments of one. The following custom hook does the job:

```
const useCount = () => {
  const [count, setCount] = useState(0);
  const inc = () => setCount((c) => c + 1);
  return [count, inc];
};
```

This opens up the entire ecosystem to provide custom hooks for various purposes. They can be a wrapper to add a tiny functionality or a huge hook that has a larger job.

You will find many custom hooks publicly available on **Node Package Manager** (**npm**) (`https://www.npmjs.com/search?q=react%20hooks`) or GitHub (`https://github.com/search?q=react+hooks&type=repositories`).

We should also discuss a little about suspense and concurrent rendering, as React hooks are designed and developed to work with these modes.

Suspense for Data Fetching and Concurrent Rendering

Suspense for Data Fetching and Concurrent Rendering are not yet released by React, but it's important to mention them briefly.

Important Note

Suspense for Data Fetching and Concurrent Rendering may have different names when they are officially released, but these are the names at the time of writing.

Suspense for Data Fetching is a mechanism that basically allows you to code your components without worrying about `async`.

Concurrent Rendering is a mechanism to split the render process into chunks to avoid blocking the **central processing unit** (**CPU**) for long periods of time.

React hooks are designed to work with these mechanisms; however, you need to avoid misusing them.

For example, one rule is that you should not mutate an existing `state` object or `ref` object. Doing so may lead to unexpected behavior such as not triggering re-renders, triggering too many re-renders, and triggering partial re-renders (meaning some components re-render while others don't when they should).

Hook functions and component functions can be invoked multiple times. Hence, another rule is those functions have to be "pure" enough so that they behave consistently, even if they are invoked several times.

These are the two major rules people often violate. This is a hard problem in practice, because even if your code violates those rules, it may just work in Non-Concurrent Rendering. Hence, people wouldn't notice the misuse. Even in Concurrent Rendering, it may work to some extent without problems, and people would only see problems occasionally. This makes it especially difficult for beginners who are using React for the first time.

Unless you are familiar with these concepts, it's better to use well-designed and battle-tested (micro) state management libraries for future/newer versions of React.

> **Important Note**
>
> As of writing, Concurrent Rendering is described in the *React 18 Working Group*, which you can read about here: `https://github.com/reactwg/react-18/discussions`.

In this section, we revisited basic React hooks and got some understanding of the concepts. Coming up, we start exploring global states, which are the main topic in this book.

Exploring global states

React provides primitive hooks such as `useState` for states that are defined in a component and consumed within the component tree. These are often called local states.

The following example uses a local state:

```
const Component = () => {
  const [state, setState] = useState();
  return (
    <div>
      {JSON.stringify(state)}
      <Child state={state} setState={setState} />
    </div>
  );
};
```

```
const Child = ({ state, setState }) => {
  const setFoo = () => setState(
    (prev) => ({ ...prev, foo: 'foo' })
  );
  return (
    <div>
      {JSON.stringify(state)}
      <button onClick={setFoo}>Set Foo</button>
    </div>
  );
};
```

On the other hand, a global state is a state that is consumed in multiple components, often far apart in an app. A global state doesn't have to be a singleton, and we may call a global state a shared state instead, to clarify that it's not a singleton.

The following code snippet provides an example of what a React component would look like with a global state:

```
const Component1 = () => {
  const [state, setState] = useGlobalState();
  return (
    <div>
      {JSON.stringify(state)}
    </div>
  );
};

const Component2 = () => {
  const [state, setState] = useGlobalState();
  return (
    <div>
      {JSON.stringify(state)}
    </div>
  );
};
```

As we haven't yet defined useGlobalState, it won't work. In this case, we want Component1 and Component2 to have the same state.

Implementing global states in React is not a trivial task. This is mostly because React is based on the component model. In the component model, locality is important, meaning a component should be isolated and should be reusable.

> **Notes about the Component Model**
> A component is a reusable piece of a unit, like a function. If you define a component, it can be used many times. This is only possible if a component definition is self-contained. If a component depends on something outside, it may not be reusable because its behavior can be inconsistent. Technically, a component itself should not depend on a global state.

React doesn't provide a direct solution for a global state, and it seems up to the developers and the community. Many solutions have been proposed, and each has its pros and cons. The goal of this book is to show typical solutions and discuss these pros and cons, which we will do in the following chapters:

- *Chapter 3, Sharing Component State with Context*
- *Chapter 4, Sharing Module State with Subscription*
- *Chapter 5, Sharing Component State with Context and Subscription*

In this section, we learned what a global state with React hooks would look like. Coming up, we will learn some basics of useState to prepare the discussion in the following chapters.

Working with useState

In this section, we will learn how to use useState, from basic usage to advanced usage. We start with the simplest form, which is updating with the state with a new value, then updating with a function, which is a very powerful feature, and finally, we will discuss lazy initialization.

Updating the state value with a value

One way to update the state value with useState is by providing a new value. You can pass a new value to the function returned by useState that will eventually replace the state value with the new value.

Here is a counter example showing updating with a value:

```
const Component = () => {
  const [count, setCount] = useState(0);
```

```
    return (
      <div>
        {count}
        <button onClick={() => setCount(1)}>
          Set Count to 1
        </button>
      </div>
    );
  };
```

You pass a value of 1 to setCount in the onClick handler. If you click the button, it will trigger Component to re-render with count=1.

What would happen if you clicked the button again? It will invoke setCount(1) again, but as it is the same value, it "bails out" and the component won't re-render. **Bailout** is a technical term in React and basically means avoiding triggering re-renders.

Let's look at another example here:

```
  const Component = () => {
    const [state, setState] = useState({ count: 0 });
    return (
      <div>
        {state.count}
        <button onClick={() => setState({ count: 1 })}>
          Set Count to 1
        </button>
      </div>
    );
  };
```

This behaves exactly the same as the previous example for the first click; however, if you click the button again, the component will re-render. You don't see any difference on screen because the count hasn't changed. This happens because the second click creates a new object, { count: 1 }, and it's different from the previous object.

Now, this leads to the following bad practice:

```
  const Component = () => {
    const [state, setState] = useState({ count: 0 });
    return (
      <div>
```

```
        {state.count}
        <button
          onClick={() => { state.count = 1; setState(state); }
        >
          Set Count to 1
        </button>
      </div>
  );
};
```

This doesn't work as expected. Even if you click the button, it won't re-render. This is because the state object is referentially unchanged, and it bails out, meaning this alone doesn't trigger the re-render.

Finally, there's an interesting usage of value update, which we can see here:

```
const Component = () => {
  const [count, setCount] = useState(0);
  return (
    <div>
      {count}
      <button onClick={() => setCount(count + 1)}>
        Set Count to {count + 1}
      </button>
    </div>
  );
};
```

Clicking the button will increment the count; however, if you click the button twice quickly enough, it will increment by just one number. This is sometimes desirable as it matches with the button title, but sometimes it's not if you expect to count how many times the button is actually clicked. That requires a function update.

Updating the state value with a function

Another way to update the state with useState is called a function update.

Here is a counter example showing updating with a function:

```
const Component = () => {
  const [count, setCount] = useState(0);
```

```
    return (
      <div>
        {count}
        <button onClick={() => setCount((c) => c + 1)}>
          Increment Count
        </button>
      </div>
    );
};
```

This actually counts how many times the button is clicked, because `(c) => c + 1` is invoked sequentially. As we saw in the previous section, value update has the same use case as the `Set Count to {count + 1}` feature. In most use cases, function updates work better if the update is based on the previous value. The `Set Count to {count + 1}` feature actually means that it doesn't depend on the previous value but depends on the displayed value.

Bailout is also possible with function updates. Here's an example to demonstrate this:

```
const Component = () => {
  const [count, setCount] = useState(0);
  useEffect(() => {
    const id = setInterval(
      () => setCount((c) => c + 1),
      1000,
    );
    return () => clearInterval(id);
  }, []);
  return (
    <div>
      {count}
      <button
        onClick={() =>
          setCount((c) => c % 2 === 0 ? c : c + 1)}
      >
        Increment Count if it makes the result even
      </button>
    </div>
```

```
  );
};
```

If the update function returns the exact same state as the previous state, it will bail out, and this component won't re-render. For example, if you invoke setCount((c) => c), it will never re-render.

Lazy initialization

useState can receive a function for initialization that will be evaluated only in the first render. We can do something like this:

```
const init = () => 0;

const Component = () => {
  const [count, setCount] = useState(init);
  return (
    <div>
      {count}
      <button onClick={() => setCount((c) => c + 1)}>
        Increment Count
      </button>
    </div>
  );
};
```

The use of init in this example is not very effective because returning 0 doesn't require much computation, but the point is that the init function can include heavy computation and is only invoked to get the initial state. The init function is evaluated lazily, not evaluated before calling useState; in other words, it's invoked just once on mount.

We have now learned how to use useState; next up is useReducer.

Using useReducer

In this section, we will learn how to use useReducer. We will learn about its typical usage, how to bail out, its usage with primitive values, and lazy initialization.

Typical usage

A reducer is helpful for complex states. Here's a simple example a with two-property object:

```
const reducer = (state, action) => {
  switch (action.type) {
    case 'INCREMENT':
      return { ...state, count: state.count + 1 };
    case 'SET_TEXT':
      return { ...state, text: action.text };
    default:
      throw new Error('unknown action type');
  }
};

const Component = () => {
  const [state, dispatch] = useReducer(
    reducer,
    { count: 0, text: 'hi' },
  );
  return (
    <div>
      {state.count}
      <button
        onClick={() => dispatch({ type: 'INCREMENT' })}
      >
        Increment count
      </button>
      <input
        value={state.text}
        onChange={(e) =>
          dispatch({ type: 'SET_TEXT', text: e.target.value })}
      />
    </div>
  );
};
```

`useReducer` allows us to define a reducer function in advance by taking the defined reducer function and initial state in parameters. The benefit of defining a reducer function outside the hook is being able to separate code and testability. Because the reducer function is a pure function, it's easier to test its behavior.

Bailout

As well as `useState`, bailout works with `useReducer` too. Using the previous example, let's modify the reducer so that it will bail out if `action.text` is empty, as follows:

```
const reducer = (state, action) => {
  switch (action.type) {
    case 'INCREMENT':
      return { ...state, count: state.count + 1 };
    case 'SET_TEXT':
      if (!action.text) {
        // bail out
        return state
      }
      return { ...state, text: action.text };
    default:
      throw new Error('unknown action type');
  }
};
```

Notice that returning `state` itself is important. If you return `{ ...state, text: action.text || state.text }` instead, it won't bail out because it's creating a new object.

Primitive value

`useReducer` works for non-object values, which are primitive values such as numbers and strings. `useReducer` with primitive values is still useful as we can define complex reducer logic outside it.

Here is a reducer example with a single number:

```
const reducer = (count, delta) => {
  if (delta < 0) {
    throw new Error('delta cannot be negative');
```

```
    }
    if (delta > 10) {
      // too big, just ignore
      return count
    }
    if (count < 100) {
      // add bonus
      return count + delta + 10
    }
    return count + delta
}
```

Notice that the action (= delta) doesn't have to have an object either. In this reducer example, the state value is a number—a primitive value—but the logic is a little more complex, with more conditions than just adding numbers.

Lazy initialization (init)

useReducer requires two parameters. The first is a reducer function and the second is an initial state. useReducer accepts an optional third parameter, which is called init, for lazy initialization.

For example, useReducer can be used like this:

```
const init = (count) => ({ count, text: 'hi' });

const reducer = (state, action) => {
  switch (action.type) {
    case 'INCREMENT':
      return { ...state, count: state.count + 1 };
    case 'SET_TEXT':
      return { ...state, text: action.text };
    default:
      throw new Error('unknown action type');
  }
};

const Component = () => {
  const [state, dispatch] = useReducer(reducer, 0, init);
```

```
  return (
    <div>
      {state.count}
      <button
        onClick={() => dispatch({ type: 'INCREMENT' })}
      >
        Increment count
      </button>
      <input
        value={state.text}
        onChange={(e) => dispatch({
          type: 'SET_TEXT',
          text: e.target.value,
        })}
      />
    </div>
  );
};
```

The init function is invoked just once on mount, so it can include heavy computation. Unlike useState, the init function takes a second argument—initialArg—in useReducer, which is 0 in the previous example.

Now we have looked at useState and useReducer separately, it's time to compare them.

Exploring the similarities and differences between useState and useReducer

In this section, we demonstrate some similarities and differences between useState and useReducer.

Implementing useState with useReducer

Implementing useState with useReducer instead is 100% possible. Actually, it's known that useState is implemented with useReducer inside React.

> **Important Note**
>
> This may not hold in the future as `useState` could be implemented more efficiently.

The following example shows how to implement `useState` with `useReducer`:

```
const useState = (initialState) => {
  const [state, dispatch] = useReducer(
    (prev, action) =>
      typeof action === 'function' ? action(prev) : action,
    initialState
  );
  return [state, dispatch];
};
```

This can then be simplified and improved upon, as follows:

```
const reducer = (prev, action) =>
  typeof action === 'function' ? action(prev) : prev;

const useState = (initialState) =>
  useReducer(reducer, initialState);
```

Here, we proved that what you can do with `useState` can be done with `useReducer`. So, wherever you have `useState`, you can just replace it with `useReducer`.

Implementing useReducer with useState

Now, let's explore if the opposite is possible—can we replace all instances of `useReducer` with `useState`? Surprisingly, it's almost true. "Almost" means there are subtle differences. But in general, people expect `useReducer` to be more flexible than `useState`, so let's see if `useState` is flexible enough in reality.

The following example illustrates how to implement the basic capability of `useReducer` with `useState`:

```
const useReducer = (reducer, initialState) => {
  const [state, setState] = useState(initialState);
  const dispatch = (action) =>
    setState(prev => reducer(prev, action));
```

```
    return [state, dispatch];
};
```

In addition to this basic capability, we can implement lazy initialization too. Let's also use `useCallback` to have a stable dispatch function, as follows:

```
const useReducer = (reducer, initialArg, init) => {
  const [state, setState] = useState(
    init ? () => init(initialArg) : initialArg,
  );
  const dispatch = useCallback(
    (action) => setState(prev => reducer(prev, action)),
    [reducer]
  );
  return [state, dispatch];
};
```

This implementation works almost perfectly as a replacement for `useReducer`. Your use case of `useReducer` is very likely handled by this implementation.

However, we have two subtle differences. As they are subtle, we don't usually consider them in too much detail. Let's learn about them in the following two subsections to get a deeper understanding.

Using the init function

One difference is that we can define `reducer` and `init` outside hooks or components. This is only possible with `useReducer` and not with `useState`.

Here is a simple count example:

```
const init = (count) => ({ count });
const reducer = (prev, delta) => prev + delta;

const ComponentWithUseReducer = ({ initialCount }) => {
  const [state, dispatch] = useReducer(
    reducer,
    initialCount,
    init
  );
```

```
  return (
    <div>
      {state}
      <button onClick={() => dispatch(1)}>+1</button>
    </div>
  );
};

const ComponentWithUseState = ({ initialCount }) => {
  const [state, setState] = useState(() =>
    init(initialCount));
  const dispatch = (delta) =>
    setState((prev) => reducer(prev, delta));
  return [state, dispatch];
};
```

As you can see in ComponentWithUseState, useState requires two inline functions, whereas ComponentWithUseReducer has no inline functions. This is a trivial thing, but some interpreters or compilers can optimize better without inline functions.

Using inline reducers

The inline reducer function can depend on outside variables. This is only possible with useReducer and not with useState. This is a special capability of useReducer.

> **Important Note**
> This capability is not usually used and not recommended unless it's really necessary.

Hence, the following code is technically possible:

```
const useScore = (bonus) =>
  useReducer((prev, delta) => prev + delta + bonus, 0);
```

This works correctly even when bonus and delta are both updated.

With the `useState` emulation, this doesn't work correctly. It would use an old `bonus` value in a previous render. This is because `useReducer` invokes the reducer function in the render phase.

As noted, this is not typically used, so overall, if we ignore this special behavior, we can say `useReducer` and `useState` are basically the same and interchangeable. You could just pick either one, based on your preference or your programming style.

Summary

In this chapter, we discussed state management and defined micro state management, in which React hooks play an important role. To prepare for the following chapters, we learned about some React hooks that are used for state management solutions, including `useState` and `useReducer`, while also looking at their similarities and differences.

In the next chapter, we learn more about a global state. For this purpose, we will discuss a local state and when a local state works, and we will then look at when a global state is required.

Part 2: Basic Approaches to the Global State

There are several approaches to using the global state effectively in React. Our focus is on optimizing re-renders. This is important because a global state can be used by multiple components. We describe three patterns – using Context, using Subscription, and using both Context and Subscription. We discuss how those patterns address the optimizing of re-renders.

This part comprises the following chapters:

2
Using Local and Global States

React components form a tree structure. In the tree structure, creating a state in a whole subtree is straightforward; you would simply create a local state in a higher component in a tree and use the state in the component and its child components. This is good in terms of locality and reusability and is why it's generally recommended to follow this strategy.

However, in some scenarios, we have a state in two or more components that are far apart in the tree. In such cases, this is where global states come in. Unlike local states, global states do not conceptually belong to a specific component, and so where we store a global state is an important point to consider.

In this chapter, we will learn about local states, including some lifting-up patterns that may be worth considering. Lifting up is a technique to put information higher in the component tree. Then, we will dive into global states and consider when to use them.

We are going to cover the following topics:

- Understanding when to use local states
- Effectively using local states
- Using global states

Technical requirements

To run the code snippets in this chapter, you need a React environment—for example, Create React App (`https://create-react-app.dev`) or CodeSandbox (`https://codesandbox.io`).

You are expected to have basic knowledge of React and React hooks, especially the concept around the component tree (`https://reactjs.org/docs/components-and-props.html`) and the `useState` hook (`https://reactjs.org/docs/hooks-reference.html#usestate`).

The code in this chapter is available on GitHub at `https://github.com/PacktPublishing/Micro-State-Management-with-React-Hooks/tree/main/chapter_02`.

Understanding when to use local states

Before we consider React, let's see how JavaScript functions work. JavaScript functions can either be pure or impure. A pure function depends only on its arguments and returns the same value as long as the arguments are the same. A state holds a value outside arguments, and functions that depend on the state become impure. React components are also functions and can be pure. If we use a state in a React component, it will be impure. However, if the state is local to the component, it doesn't affect other components, and we call this characteristic "contained."

In this section, we learn JavaScript functions, and how similar React components are to JavaScript functions. We then discuss how a local state is conceptually implemented.

Functions and arguments

In JavaScript, a function takes an argument and returns a value. For example, here's a simple function:

```
const addOne = (n) => n + 1;
```

This is a pure function that always returns the same value for the same argument. It is often the case that pure functions are preferred because their behavior is predictable.

A function can depend on global variables, such as the following:

```
let base = 1;

const addBase = (n) => n + base;
```

The `addBase` function works exactly the same as `addOne`, as long as `base` isn't changed. However, if at some point we change `base` to `base=2`, it behaves differently. This is not a bad thing at all, and it's actually a powerful feature as you can change the function behavior from outside. The downside is that you can't simply grab the `addBase` function and use it arbitrarily somewhere else without knowing it depends on an outside variable. As you can tell, it's a trade-off.

This is not a preferred pattern if `base` is a **singleton** (a single value in memory) because the code becomes less reusable. To avoid the singleton and mitigate the downside a little, a more modular approach would be to create a container object, as follows:

```
const createContainer = () => {
  let base = 1;
  const addBase = (n) => n + base;
  const changeBase = (b) => { base = b; };
  return { addBase, changeBase };
};


const { addBase, changeBase } = createContainer();
```

This is no longer a singleton, and you can create as many containers as you want. Unlike having a `base` global variable as a singleton, containers are isolated and are more reusable. You can use a container in one part of your code without affecting other parts of your code with a different container.

A small note: although `addBase` in a container is not a mathematically pure function, you can get the same result by calling `addBase` if `base` is not changed (this characteristic is sometimes called **idempotent**).

React components and props

React is conceptually a function that converts a state to a **user interface** (**UI**). When you code with React, the React component is literally a JavaScript function, and its arguments are called props.

A function component that shows a number will look like this:

```
const Component = ({ number }) => {
  return <div>{number}</div>;
};
```

This component takes a `number` argument and returns a **JavaScript syntax extension (JSX)** element that represents the `number` on screen.

> **What Is a JSX Element?**
> JSX is a syntax with angle brackets to produce React elements. A React element is a data structure to represent a part of the UI. We may refer to React elements as JSX elements, especially when React elements are in JSX syntax.

Now, let's make another component that shows a `number + 1`, as follows:

```
const AddOne = ({ number }) => {
  return <div>{number + 1}</div>;
};
```

This component takes `number` and returns the `number + 1`. This behaves exactly like `addOne` in the previous section, and this is a pure function. The only differences are that the argument is a props object and the return value is in JSX format.

Understanding useState for local states

What if we use `useState` for a local state? Let's make `base` a state and display a `number` that we can add to it, as follows:

```
const AddBase = ({ number }) => {
  const [base, changeBase] = useState(1);
  return <div>{number + base}</div>;
};
```

This function is not technically pure as it depends on `base`, which is not in the function arguments.

What does `useState` in `AddBase` do? Let's remind ourselves of `createContainer` in the previous section. As `createContainer` returns `base` and `changeBase`, `useState` returns `base` and `changeBase` in a tuple (meaning a structure of two or more values—in this case, two). We don't explicitly see how `base` and `changeBase` are created in this code, but it's conceptually similar.

If we assume the `useState` behavior, meaning it returns `base` unless changed, the `AddBase` function is idempotent, as we saw with `createContainer`.

This `AddBase` function with `useState` is contained because `changeBase` is only available within the scope of the function declaration. It's impossible to change `base` outside the function. This usage of `useState` is a local state, and because it's contained and doesn't affect anything outside the component, it ensures locality; this usage is preferred whenever appropriate.

Limitation of local states

When is a local state not appropriate? It isn't appropriate when we want to break the locality. In the `AddBase` component example, it's when we want to change `base` from a totally different part of the code. If you need to change `state` from outside the function component, that's when a global state comes in.

The state variable is conceptually a global variable. A global variable is useful to control a JavaScript function's behavior from outside the function. Likewise, a global state is useful to control React component behavior from outside the component. However, using a global state makes the component behavior less predictable. It's a trade-off. We shouldn't use global states more than we need to. Consider using local states as a primary means and only use global states for a secondary mean. In this sense, it's important to learn how many use cases local states can cover.

In this section, we learned about a local state in React, alongside JavaScript functions. Coming up, we will learn some patterns to use local states.

Effectively using local states

There are some patterns you should know to be able to use a local state effectively. In this section, we will learn how to lift states up, which means defining a state higher in the component tree, and lifting content up, which means defining a content higher in the component tree.

Lifting state up

Let's suppose we have two counter components, as follows:

```
const Component1 = () => {
  const [count, setCount] = useState(0);
  return (
    <div>
      {count}
      <button onClick={() => setCount((c) => c + 1)}>
        Increment Count
```

```
      </button>
    </div>
  );
};

const Component2 = () => {
  const [count, setCount] = useState(0);
  return (
    <div>
      {count}
      <button onClick={() => setCount((c) => c + 1)}>
        Increment Count
      </button>
    </div>
  );
};
```

Because there are two separate local states defined in the two components, these two counters work separately. In case we want to share the state and make it work for a single shared counter, we can create a parent component and lift the state up.

Here is an example with a single parent component that contains both Component1 and Component2 as children and passes props to them:

```
const Component1 = ({ count, setCount }) => {
  return (
    <div>
      {count}
      <button onClick={() => setCount((c) => c + 1)}>
        Increment Count
      </button>
    </div>
  );
};

const Component2 = ({ count, setCount }) => {
  return (
    <div>
```

```
      {count}
      <button onClick={() => setCount((c) => c + 1)}>
        Increment Count
      </button>
    </div>
  );
};

const Parent = () => {
  const [count, setCount] = useState(0);
  return (
    <>
      <Component1 count={count} setCount={setCount} />
      <Component2 count={count} setCount={setCount} />
    </>
  );
};
```

Because the count state is defined just once in `Parent`, the state is shared between `Component1` and `Component2`. This is still a local state in a component; its child components can use the state from the parent component.

This pattern would work in most use cases with a local state; however, there's a slight concern about performance. If we lift up the state, `Parent` will render as well as the entire subtree, including all its child components. This may be a performance issue in some use cases.

Lift content up

With complex component trees, we may have a component that doesn't depend on the state we are lifting up.

In the following example, we add a new `AdditionalInfo` component to `Component1` from the previous example:

```
const AdditionalInfo = () => {
  return <p>Some information</p>
};

const Component1 = ({ count, setCount }) => {
```

```
    return (
      <div>
        {count}
        <button onClick={() => setCount((c) => c + 1)}>
          Increment Count
        </button>
        <AdditionalInfo />
      </div>
    );
};

const Parent = () => {
  const [count, setCount] = useState(0);
  return (
    <>
      <Component1 count={count} setCount={setCount} />
      <Component2 count={count} setCount={setCount} />
    </>
  );
};
```

If the count is changed, the `Parent` re-renders, and then `Component1`, `Component2`, and `AdditionalInfo` re-render too. However, `AdditionalInfo` doesn't have to re-render in this case because it doesn't depend on `count`. This is an extra re-render that should be avoided if it has an impact on performance.

To avoid extra re-renders, we can lift up content. In this case, `Parent` re-renders with `count`, hence, we create `GrandParent`, as follows:

```
const AdditionalInfo = () => {
  return <p>Some information</p>
};

const Component1 = ({ count, setCount, additionalInfo }) => {
  return (
    <div>
      {count}
      <button onClick={() => setCount((c) => c + 1)}>
```

```
          Increment Count
        </button>
        {additionalInfo}
      </div>
    );
  };


  const Parent = ({ additionalInfo }) => {
    const [count, setCount] = useState(0);
    return (
      <>
        <Component1
          count={count}
          setCount={setCount}
          additionalInfo={additionalInfo}
        />
        <Component2 count={count} setCount={setCount} />
      </>
    );
  };


  const GrandParent = () => {
    return <Parent additionalInfo={<AdditionalInfo />} />;
  };
```

The GrandParent component has additionalInfo (a JSX element), which is passed down to the children. By doing this, AdditionalInfo doesn't re-render when count changes. This is a technique we should consider not only for performance but also for organizing your component tree structure.

A variant of this is to use children props. The following example using children props is equivalent to the previous example, but with a different coding style:

```
  const AdditionalInfo = () => {
    return <p>Some information</p>
  };


  const Component1 = ({ count, setCount, children }) => {
```

```
    return (
      <div>
        {count}
        <button onClick={() => setCount((c) => c + 1)}>
          Increment Count
        </button>
        {children}
      </div>
    );
};

const Parent = ({ children }) => {
  const [count, setCount] = useState(0);
  return (
    <>
      <Component1 count={count} setCount={setCount}>
        {children}
      </Component1>
      <Component2 count={count} setCount={setCount} />
    </>
  );
};

const GrandParent = () => {
  return (
    <Parent>
      <AdditionalInfo />
    </Parent>
  );
};
```

children is a special prop name that is represented as nested children elements in JSX format. If you have several elements to pass, naming your props would fit better. It's mostly a stylistic choice, and developers can take whichever approach they prefer.

In this section, we learned some patterns to effectively use local states. If we lift up states and content properly, we should be able to solve various use cases with only local states. Coming up, we will learn how to use global states.

Using global states

In this section, we will learn what a global state is again and when we should use it.

What is a global state?

In this book, a global state simply means that it's *not* a local state. If a state conceptually belongs to a single component and is encapsulated by the component, it is a local state. Hence, if a state doesn't belong to a single component and can be used by multiple components, it is a global state.

There could be an application-wide local state that all components depend on. In this case, the application-wide local state can be seen as a global state. In this sense, we can't clearly divide local states and global states. In most cases, if you consider where a state conceptually belongs, you can work out whether it's local or global.

There are two aspects when people talk about a global state, as outlined here:

- One is a singleton, meaning that in some contexts, the state has one value.
- The other is a shared state, which means that the state value is shared among different components, but it doesn't have to be the single value in JavaScript memory. A global state that is not a singleton can have multiple values.

To illustrate how a non-singleton global state works, here is an example to show a non-singleton variable in JavaScript:

```
const createContainer = () => {
  let base = 1;
  const addBase = (n) => n + base;
  const changeBase = (b) => { base = b; };
  return { addBase, changeBase };
};

const container1 = createContainer();
const container2 = createContainer();

container1.changeBase(10);

console.log(container1.addBase(2)); // shows "3"
console.log(container2.addBase(2)); // shows "12"
```

In this example, base is a scoped variable in a container. As base is isolated in each container, changing base in container1 doesn't affect base in container2.

In React, the concept is similar. If a global state is a singleton, we have only one value in memory. If a global state is non-singleton, we may have multiple values for different parts (subtrees) of a component tree.

When to use global states

There are two guidelines for when we need a global state in React, as follows:

- When passing a prop is not desirable
- When we already have a state outside of React

Let's discuss each of them.

Prop passing is not desirable

If you need a state in two components that are far away in the component tree, it would not be desirable to have a state in the common root component and then pass the state all the way down to the two components.

For example, if our tree is three levels deep and we need to lift up the state to the top, it would look like this:

```
const Component1 = ({ count, setCount }) => {
  return (
    <div>
      {count}
      <button onClick={() => setCount((c) => c + 1)}>
        Increment Count
      </button>
    </div>
  );
};

const Parent = ({ count, setCount }) => {
  return (
    <>
      <Component1 count={count} setCount={setCount} />
    </>
```

```
    );
};

const GrandParent = ({ count, setCount }) => {
  return (
    <>
      <Parent count={count} setCount={setCount} />
    </>
  );
};

const Root = () => {
  const [count, setCount] = useState(0);
  return (
    <>
      <GrandParent count={count} setCount={setCount} />
    </>
  );
};
```

This is totally fine and recommended for locality; however, it could be too tedious to have your intermediate components used to pass props. Passing props through multi-level intermediate components might not result in a good developer experience, because it could seem like unnecessary extra work. Furthermore, the intermediate components re-render when the state is updated, which may impact performance.

In such cases, having a global state is more appropriate, and no intermediate components need to take care of passing the state.

Here is some pseudo code showing how a global state would work with the previous example:

```
const Component1 = () => {
  // useGlobalCountState is a pseudo hook
  const [count, setCount] = useGlobalCountState();
  return (
    <div>
      {count}
      <button onClick={() => setCount((c) => c + 1)}>
```

```
            Increment Count
          </button>
        </div>
      );
    };

    const Parent = () => {
      return (
        <>
          <Component1 />
        </>
      );
    };

    const GrandParent = () => {
      return (
        <>
          <Parent />
        </>
      );
    };

    const Root = () => {
      return (
        <>
          <GrandParent />
        </>
      );
    };
```

In this example, the only component that uses a global state is Component1. Unlike with local states and prop passing, no intermediate components, Parent and GrandParent, know about a global state.

Already have a state outside of React

In some cases, you would already have a global state outside of React, as having a global state outside is more straightforward. For example, your app might have user-authenticated information that you obtained without React somehow. In such an example, a global state should exist outside React, and the authentication information could be stored in a global state.

Here is some pseudo code showing such an example:

```
const globalState = {
  authInfo: { name: 'React' },
};

const Component1 = () => {
  // useGlobalState is a pseudo hook
  const { authInfo } = useGlobalState();
  return (
    <div>
      {authInfo.name}
    </div>
  );
};
```

In this example, `globalState` exists and is defined outside React. `useGlobalState` is a hook that would connect to `globalState` and that could provide `authInfo` in `Component1`.

In this section, we learned that a global state is a state that can't be a local state. Global state is mainly used secondary to local states, and there are two patterns where using a global state works well: one is in a case where prop passing doesn't make sense, and the other is where a global state already exists in an app.

Summary

In this chapter, we discussed local states and global states. Local states are preferable whenever possible, and we learned some techniques to use local states effectively. However, global states play a role where local states do not, which is why we looked at when you should use global states instead.

In the next three chapters, we will learn three patterns to implement a global state in React; in the next chapter specifically, we will start with utilizing React context.

3
Sharing Component State with Context

React has provided Context since version 16.3. Context has nothing to do with states, but it's a mechanism for passing data from component to component instead of using props. By combining Context with a component state, we can provide a global state.

In addition to the Context support provided since React 16.3, React 16.8 introduced the `useContext` hook. By using `useContext` and `useState` (or `useReducer`), we can create custom hooks for a global state.

Context is not fully designed for global states. One of the known limitations is that all Context consumers re-render upon updates, which can lead to extra re-renders. It's a general recommendation to split a global state into pieces.

In this chapter, we discuss the general recommendations for using Context and show some concrete examples. We also discuss some techniques to use Context with TypeScript. The goal is to make you feel confident with using Context for a global state.

In this chapter, we will cover the following topics:

- Exploring `useState` and `useContext`
- Understanding Context

- Creating a Context for a global state
- Best practices for using Context

Technical requirements

If you are new to React Context, it's highly recommended to learn some basics; check out the official documentation (`https://reactjs.org/docs/context.html`) and the official blog (`https://reactjs.org/blog/2018/03/29/react-v-16-3.html`).

You are also expected to have general knowledge around React including React hooks; you can refer to the official site (`https://reactjs.org`) to learn more.

In some code, we use TypeScript, which you should have basic knowledge of; you can find out more here: `https://www.typescriptlang.org`.

The code in this chapter is available on GitHub at `https://github.com/PacktPublishing/Micro-State-Management-with-React-Hooks/tree/main/chapter_03`.

To run the code snippets in this chapter, you need a React environment—for example, Create React App (`https://create-react-app.dev`) or CodeSandbox (`https://codesandbox.io`).

Exploring useState and useContext

By combining `useState` and `useContext`, we can create a simple global state. Let's recap on how to use `useState` without `useContext`, how `useContext` works for static values, and how we combine `useState` and `useContext`.

Using useState without useContext

Before diving into `useContext`, let's be reminded of how to `useState`, with a concrete example. This example is going to be a reference for the following examples in the chapter.

Here, we define a `count` state with `useState` higher in the component tree and pass the state value and the update function down the tree.

In the `App` component, we use `useState` and get `count` and `setCount`, which are passed to the `Parent` component. The code is illustrated in the following snippet:

```
const App = () => {
  const [count, setCount] = useState(0);
```

```
    return <Parent count={count} setCount={setCount} />;
};
```

This is a very basic pattern, which we know as *lifting the state up*, from *Chapter 2, Using Local and Global States.*

Now, let's define a `Parent` component. It passes the two props to `Component1` and `Component2`, as follows:

```
const Parent = ({ count, setCount }) => (
    <>
        <Component1 count={count} setCount={setCount} />
        <Component2 count={count} setCount={setCount} />
    </>
);
```

This passing of props from parent to children is a repetitive task and is often referred to as **prop drilling**.

`Component1` and `Component2` display the `count` state and a button to increase the `count` state with `setCount`, as illustrated in the following code snippet:

```
const Component1 = ({ count, setCount }) => (
    <div>
        {count}
        <button onClick={() => setCount((c) => c + 1)}>
            +1
        </button>
    </div>
);

const Component2 = ({ count, setCount }) => (
    <div>
        {count}
        <button onClick={() => setCount((c) => c + 2)}>
            +2
        </button>
    </div>
);
```

These two components are pure components, which means they receive props and display things based only on those props. `Component2` is slightly different from `Component1`, which increases the count by two. If it were identical, we wouldn't need to define two components.

There is nothing wrong with this example. Only when the app gets bigger, passing props down the tree, will this not make sense. In this case, the `Parent` component doesn't necessarily need to know about the `count` state, and it may make sense to hide the existence of the `count` state in the `Parent` component.

Using useContext with a static value

React Context helps to eliminate props. It's a means to pass a value from a parent component to its children under the tree, without using props.

The following example shows how to use React Context with a static value. It has multiple providers to provide different values. Providers can be nested, and a consumer component (a consumer component means a component with `useContext`) will pick the closest provider in the component tree to get the Context value. There is only one component with `useContext` to consume the Context, and the component is used in multiple places.

Firstly, we define a color Context with `createContext`, which takes a default value, as follows:

```
const ColorContext = createContext('black');
```

In this case, the default value for the color Context is `'black'`. The default value is used if a component is not in any providers.

Now, we define a consumer component. It reads the color Context and displays a text in that color. The code is illustrated in the following snippet:

```
const Component = () => {
  const color = useContext(ColorContext);
  return <div style={{ color }}>Hello {color}</div>;
};
```

`Component` reads the `color` Context value, but at this point, we don't know what the color is, and it literally depends on the Context.

Finally, we define an App component. The component tree in the App component has multiple ColorContext.Provider components with different colors. The code is illustrated in the following snippet:

```
const App = () => (
  <>
    <Component />
    <ColorContext.Provider value="red">
      <Component />
    </ColorContext.Provider>
    <ColorContext.Provider value="green">
      <Component />
    </ColorContext.Provider>
    <ColorContext.Provider value="blue">
      <Component />
      <ColorContext.Provider value="skyblue">
        <Component />
      </ColorContext.Provider>
    </ColorContext.Provider>
  </>
);
```

The first Component instance shows the color "black" because it's not wrapped by any providers. The second and the third show "red" and "green" respectively. The fourth Component instance shows "blue", and the last Component instance shows "skyblue", because the closest provider has the value "skyblue" even though it's inside the provider with "blue".

Multiple providers and reusing the consumer component is an important capability of React Context. If this capability is not important for your use case, you might not need React Context. We will discuss the subscription method without Context in *Chapter 4, Sharing Module State with Subscription.*

Using useState with useContext

Now, let's learn how the combination of useState and useContext structure our code. We can pass the state value and update function in Context instead of props.

The following example implements a simple count state with useState and useContext. We define a Context that holds both the count state value and the setCount update function. The Parent component doesn't take props, and Component1 and Component2 use useContext to get the state.

First, we create a Context for the count state. The default value holds a static count value and a fallback empty setCount function. The code is illustrated in the following snippet:

```
const CountStateContext = createContext({
  count: 0,
  setCount: () => {},
});
```

The default value helps to infer types in TypeScript. However, in most cases, we need a state instead of a static value, as the default value is not very useful. Using the default value is almost unintentional in such cases, so we may throw an error instead. We will discuss some best practices later in the *Best practices for using Context* section.

The App component has a state with useState, and passes count and setCount to the created Context provider component, as illustrated in the following code snippet:

```
const App = () => {
  const [count, setCount] = useState(0);
  return (
    <CountStateContext.Provider
      value={{ count, setCount }}
    >
      <Parent />
    </CountStateContext.Provider>
  );
};
```

The Context value we pass to CountStateContext.Provider is an object containing count and setCount. This object has the same structure as the default value.

We define a `Parent` component. Unlike the example in the previous section, we don't need to pass props. The code is illustrated in the following snippet:

```
const Parent = () => (
  <>
    <Component1 />
    <Component2 />
  </>
);
```

Even though the `Parent` component is in the Context provider in App, it does not know about the existence of the `count` state. The components inside `Parent` can still use the `count` state through the Context.

Finally, we define `Component1` and `Component2`. They take `count` and `setCount` from the Context value instead of props. The code is illustrated in the following snippet:

```
const Component1 = () => {
  const { count, setCount } =
    useContext(CountStateContext);
  return (
    <div>
      {count}
      <button onClick={() => setCount((c) => c + 1)}>
        +1
      </button>
    </div>
  );
};

const Component2 = () => {
  const { count, setCount } =
    useContext(CountStateContext);
  return (
    <div>
      {count}
      <button onClick={() => setCount((c) => c + 2)}>
        +2
      </button>
```

```
      </div>
   );
};
```

What is the Context value these components get? They get the Context value from the closest provider. We can use multiple providers to provide isolated count states, which, again, is an important capability of using React Context.

In this section, we learned about React Context and how to create a simple global state with it. Coming up, we will dive into React Context behavior.

Understanding Context

When a Context provider has a new Context value, all Context consumers receive the new value and re-render. This means the value in the provider is propagated to all the consumers. It is important for us to understand how Context propagation works and its limitations.

How Context propagation works

If you use a Context provider, you can update the Context value. When a Context provider receives a new Context value, it triggers *all* the Context consumer components to re-render.

It's sometimes the case that a child component re-renders for two reasons—one because of the parent, and the other because of the Context.

To stop re-rendering without Context value changes, in this case, we can use the *lift content up* technique, or memo. memo is a function to wrap a component and is used to prevent re-renders if the component props don't change.

Let's see an example with some components wrapped with memo to understand its behavior.

As with previous examples, we again use a simple Context that holds a color string, as follows:

```
const ColorContext = createContext('black');
```

'black' is the default value, which will be used if there are no Context providers found in the component tree.

We then define `ColorComponent`, which is similar to previous examples, but it also has `renderCount` to show how many times this component is rendered, as illustrated in the following code snippet:

```
const ColorComponent = () => {
  const color = useContext(ColorContext);
  const renderCount = useRef(1);
  useEffect(() => {
    renderCount.current += 1;
  });
  return (
    <div style={{ color }}>
      Hello {color} (renders: {renderCount.current})
    </div>
  );
};
```

We use `useRef` for `renderCount`. `renderCount.current` is a number indicating the render count. The `renderCount.current` number is incremented by one with `useEffect`.

Next is `MemoedColorComponent`, which is `ColorComponent` wrapped by memo. The code is illustrated in the following snippet:

```
const MemoedColorComponent = memo(ColorComponent);
```

The memo function is to create a memoized component from a base component. The memoized component produces a stable result for the same props.

We define another component, `DummyComponent`, which doesn't use `useContext`. The code is illustrated in the following snippet:

```
const DummyComponent = () => {
  const renderCount = useRef(1);
  useEffect(() => {
    renderCount.current += 1;
  });
  return <div>Dummy (renders: {renderCount.current})</div>;
};
```

This component is to compare the behavior against `ColorComponent`.

We also define `MemoedDummyComponent` for `DummyComponent` with `memo`, as follows:

```
const MemoedDummyComponent = memo(DummyComponent);
```

Next, we define a `Parent` component; it has four kinds of components we defined previously. The code is illustrated in the following snippet:

```
const Parent = () => (
  <ul>
    <li><DummyComponent /></li>
    <li><MemoedDummyComponent /></li>
    <li><ColorComponent /></li>
    <li><MemoedColorComponent /></li>
  </ul>
);
```

Finally, the `App` component has a color state with `useState` and passes the value to `ColorContext.Provider`. It also shows a text field to change the color state. The code is illustrated in the following snippet:

```
const App = () => {
  const [color, setColor] = useState('red');
  return (
    <ColorContext.Provider value={color}>
      <input
        value={color}
        onChange={(e) => setColor(e.target.value)}
      />
      <Parent />
    </ColorContext.Provider>
  );
};
```

This example behaves in the following way:

1. Initially, all the components render.

2. If you change the value in the text input, the `App` component renders because of `useState`.

3. Then, `ColorContext.Provider` gets a new value, and at the same time, the `Parent` component renders.

4. `DummyComponent` renders but `MemoedDummyComponent` doesn't.

5. `ColorComponent` renders for two reasons—firstly, the parent renders, and secondly, the Context changes.

6. `MemoedColorComponent` renders because the Context changes.

What's important to learn here is that `memo` doesn't stop the internal Context consumer from re-rendering. This is obviously unavoidable because otherwise, components could have inconsistent Context values.

Limitations when using Context for objects

Using primitive values for Context values is intuitive, but using object values may require caution due to their behavior. An object may contain several values, and Context consumers may not use all of them.

The following example is to reproduce such a case where a component uses only part of an object.

First, we define a Context whose value is an object with two counts, `count1` and `count2`, as follows:

```
const CountContext = createContext({ count1: 0, count2: 0 });
```

Using this count Context, we define a `Counter1` component is to show `count1`. We have `renderCount` to show the render count. We also define a `MemoedCounter1` component, which is the memoized component. The code is illustrated in the following snippet:

```
const Counter1 = () => {
  const { count1 } = useContext(CountContext);
  const renderCount = useRef(1);
  useEffect(() => {
    renderCount.current += 1;
  });
```

```
  return (
    <div>
      Count1: {count1} (renders: {renderCount.current})
    </div>
  );
};


const MemoedCounter1 = memo(Counter1);
```

Notice that the `Counter1` component uses only `count1` from the Context value.

Likewise, we define a `Counter2` component that shows `count2` and the memoized `MemoCounter2` component, as follows:

```
const Counter2 = () => {
  const { count2 } = useContext(CountContext);
  const renderCount = useRef(1);
  useEffect(() => {
    renderCount.current += 1;
  });
  return (
    <div>
      Count2: {count2} (renders: {renderCount.current})
    </div>
  );
};


const MemoCounter2 = memo(Counter2);
```

The `Parent` component has two memoized components, as illustrated in the following code snippet:

```
const Parent = () => (
  <>
    <MemoCounter1 />
    <MemoCounter2 />
  </>
);
```

Finally, the App component has two counts with two useState hooks and provides the two counts with one Context. It has two buttons to increment two counts respectively, as illustrated in the following code snippet:

```
const App = () => {
  const [count1, setCount1] = useState(0);
  const [count2, setCount2] = useState(0);
  return (
    <CountContext.Provider value={{ count1, count2 }}>
      <button onClick={() => setCount1((c) => c + 1)}>
        {count1}
      </button>
      <button onClick={() => setCount2((c) => c + 1)}>
        {count2}
      </button>
      <Parent />
    </CountContext.Provider>
  );
};
```

Notice again that the place of two buttons is not very important.

The two counts, count1 and count2, are totally separate—Counter1 uses only count1 and Counter2 uses only count2. Hence, ideally, Counter1 should re-render only when count1 is changed. If Counter1 re-renders without changing count1, it produces the same result, which means that was just an extra re-render. In this example, Counter1 re-renders even when only count2 is changed.

This is the extra re-render limitation in the behavior that we should be aware of when we utilize React Context.

> **Extra Re-Renders**
>
> Extra re-renders are a pure overhead that should be technically avoided. However, this would be fine unless performance is a big concern because users wouldn't notice a few extra re-renders. Overengineering to avoid a few extra re-renders might not be worth resolving in practice.

In this section, we learned about the behavior of React Context and why it's limited to being used with objects. Coming up, we learn some typical patterns for implementing a global state with Context.

Creating a Context for a global state

Based on the React Context behavior, we will discuss two solutions regarding using Context with a global state, as follows:

- Creating small state pieces
- Creating one state with useReducer and propagating with multiple Contexts

Let's take a look at each solution.

Creating small state pieces

The first solution is to split a global state into pieces. So, instead of using a big combined object, create a global state and a Context for each piece.

The following example creates two count states, with a Context and a provider component for each count state.

Firstly, we define two Contexts, Count1Context and Count2Context, one for each piece, as follows:

```
type CountContextType = [
  number,
  Dispatch<SetStateAction<number>>
];

const Count1Context = createContext<CountContextType>([
  0,
  () => {}
]);
const Count2Context = createContext<CountContextType>([
  0,
  () => {}
]);
```

The Context value is a tuple of a count value and an updating function. We specified a static value and a dummy function as a default value.

We then define a `Counter1` component that only uses `Count1Context`, as follows:

```
const Counter1 = () => {
  const [count1, setCount1] = useContext(Count1Context);
  return (
    <div>
      Count1: {count1}
      <button onClick={() => setCount1((c) => c + 1)}>
        +1
      </button>
    </div>
  );
};
```

Notice the implementation of `Counter1` only depends on `Count1Context`, and it doesn't know about any other Contexts.

Likewise, we define a `Counter2` component that uses only `Count2Context`, as follows:

```
const Counter2 = () => {
  const [count2, setCount2] = useContext(Count2Context);
  return (
    <div>
      Count2: {count2}
      <button onClick={() => setCount2((c) => c + 1)}>
        +1
      </button>
    </div>
  );
};
```

The `Parent` component has `Counter1` and `Counter2` components, as illustrated in the following code snippet:

```
const Parent = () => (
  <div>
    <Counter1 />
    <Counter1 />
    <Counter2 />
```

```
      <Counter2 />
    </div>
  );
```

The `Parent` component has two counters each, just for demonstration purposes.

We define a `Count1Provider` component for `Count1Context`. The
`Count1Provider` component has a `count` state with `useState` and passes the
count value and `update` function to the `Count1Context.Provider` component,
as illustrated in the following code snippet:

```
const Count1Provider = ({
  children
}: {
  children: ReactNode
}) => {
  const [count1, setCount1] = useState(0);
  return (
    <Count1Context.Provider value={[count1, setCount1]}>
      {children}
    </Count1Context.Provider>
  );
};
```

Likewise, we define a `Count2Provider` component for `Count2Context`, as follows:

```
const Count2Provider = ({
  children
}: {
  children: ReactNode
}) => {
  const [count2, setCount2] = useState(0);
  return (
    <Count2Context.Provider value={[count2, setCount2]}>
      {children}
    </Count2Context.Provider>
  );
};
```

The `Count1Provider` and `Count2Provider` components are similar; the only difference is the Context to provide a value.

Finally, the `App` component has a `Parent` component with two provider components, as illustrated in the following code snippet:

```
const App = () => (
  <Count1Provider>
    <Count2Provider>
      <Parent />
    </Count2Provider>
  </Count1Provider>
);
```

Notice the `App` component has two provider components nested. Having more provider components lead to deeper nesting. We will discuss mitigating nesting in the *Best practices for using Context* section.

This example doesn't suffer from the extra re-render limitation we described in the previous section. This is because Contexts hold only primitive values. The `Counter1` and `Counter2` components only re-render when `count1` and `count2` are changed respectively. It is necessary to create a provider for each state; otherwise, `useState` would return a new tuple object and a Context would trigger re-renders.

If you are sure that an object is used at once and the usage doesn't hit the limitation of the Context behavior, putting an object as a Context value is totally acceptable. Here's an example of a `user` object that would be used at once:

```
const [user, setUser] = useState({
  firstName: 'react',
  lastName: 'hooks'
});
```

In this case, it doesn't make sense to split it into Contexts. Using a single Context for a `user` object would be better.

Next, let's look at another solution.

Creating one state with useReducer and propagating with multiple Contexts

The second solution is to create a single state and use multiple Contexts to distribute state pieces. In this case, distributing a function to update the state should be done with a separate Context.

The following example is based on `useReducer`. It has three Contexts; two are for state pieces, and the last one is for a dispatch function.

First, we create two value Contexts for two counts, and one Context for the dispatch function that will be used to update the two counts, as follows:

```
type Action = { type: "INC1" } | { type: "INC2" };

const Count1Context = createContext<number>(0);
const Count2Context = createContext<number>(0);
const DispatchContext = createContext<Dispatch<Action>>(
  () => {}
);
```

In this case, if we have more counts, we create more count Contexts, but the dispatch Context will remain just one.

We define a reducer for the dispatch function later in this example.

Next, we define a `Counter1` component that uses two Contexts—one for the value and another for the dispatch function, as follows:

```
const Counter1 = () => {
  const count1 = useContext(Count1Context);
  const dispatch = useContext(DispatchContext);
  return (
    <div>
      Count1: {count1}
      <button onClick={() => dispatch({ type: "INC1" })}>
        +1
      </button>
    </div>
  );
};
```

The `Counter1` component reads `count1` from `Count1Context`.

We define a `Counter2` component, which is just like `Counter1` except that it reads `count2` from a different Context. The code is illustrated in the following snippet:

```
const Counter2 = () => {
  const count2 = useContext(Count2Context);
  const dispatch = useContext(DispatchContext);
  return (
    <div>
      Count2: {count2}
      <button onClick={() => dispatch({ type: "INC2" })}>
        +1
      </button>
    </div>
  );
};
```

Both `Counter1` and `Counter2` components use the same `DispatchContext` Context.

The `Parent` component is the same as the previous example, as we can see here:

```
const Parent = () => (
  <div>
    <Counter1 />
    <Counter1 />
    <Counter2 />
    <Counter2 />
  </div>
);
```

Now, we define a `Provider` component that is unique in this example. The `Provider` component uses `useReducer`. The reducer function handles two action types—`INC1` and `INC2`. The `Provider` component includes providers from three Contexts that we defined previously. The code is illustrated in the following snippet:

```
const Provider = ({ children }: { children: ReactNode }) => {
  const [state, dispatch] = useReducer(
    (
      prev: { count1: number; count2: number },
```

```
      action: Action
  ) => {
    if (action.type === "INC1") {
      return { ...prev, count1: prev.count1 + 1 };
    }
    if (action.type === "INC2") {
      return { ...prev, count2: prev.count2 + 1 };
    }
    throw new Error("no matching action");
  },
  {
    count1: 0,
    count2: 0,
  }
);
return (
  <DispatchContext.Provider value={dispatch}>
    <Count1Context.Provider value={state.count1}>
      <Count2Context.Provider value={state.count2}>
        {children}
      </Count2Context.Provider>
    </Count1Context.Provider>
  </DispatchContext.Provider>
);
};
```

The code is a bit long because of the reducer, which can be more complex. The point is nested providers, providing each state piece and one dispatch function.

Finally, the App component just has the Provider component and the Parent component in it, as illustrated in the following code snippet:

```
const App = () => (
  <Provider>
    <Parent />
  </Provider>
);
```

This example also doesn't suffer from the extra re-render limitation; changing `count1` in the state only triggers `Counter1` to re-render, while `Counter2` is not affected.

The benefit of using a single state over using multiple states in the previous example is that the single state can update multiple pieces with one action. For example, you can add something like this in the reducer:

```
if (action.type === "INC_BOTH") {
  return {
    ...prev,
    count1: prev.count1 + 1,
    count2: prev.count2 + 1,
  };
}
```

As we discussed in the first solution, it is acceptable to create a Context for an object (such as the `user` object) in this solution too.

In this section, we learned two solutions to use Context for a global state. They are typical solutions, but there would be many variants. The key point is to use multiple Contexts to avoid extra re-renders. In the next section, we learn some best practices to deal with a global state based on multiple Contexts.

Best practices for using Context

In this section, we will learn three patterns to deal with Contexts for a global state, as follows:

- Creating custom hooks and provider components
- A factory pattern with a custom hook
- Avoiding provider nesting with `reduceRight`

Let's take a look at each one.

Creating custom hooks and provider components

In the previous examples in this chapter, we directly used `useContext` to get Context values. Now, we will explicitly create custom hooks to access Context values as well as provider components. This allows us to hide Contexts and restrict their usage.

The following example creates custom hooks and provider components. We make a default Context value `null` and check if the value is `null` in the custom hooks. This checks if the custom hooks are used under the providers.

The first thing we do, as always, is to create a Context; this time, the default value of the Context is `null`, which indicates that the default value can't be used and the provider is always is required. The code is illustrated in the following snippet:

```
type CountContextType = [
  number,
  Dispatch<SetStateAction<number>>
];


const Count1Context = createContext<
  CountContextType | null
>(null);
```

We then define `Count1Provider`, which creates a state with `useState` and passes it to `Count1Context.Provider`, as illustrated in the following code snippet:

```
export const Count1Provider = ({
  children
}: {
  children: ReactNode
}) => (
  <Count1Context.Provider value={useState(0)}>
    {children}
  </Count1Context.Provider>
);
```

Notice that we use `useState(0)` in the **JavaScript syntax extension (JSX)** form. This is valid, and it's short for having `const [count, setCount] = useState(0);` and `return <Count1Context.Provider value={[count, setCount]}>` in one line.

Next, we define a `useCount1` hook to return a value from `Count1Context`. Here, we check that `null` from the Context value throws a meaningful error. Developers often make mistakes, and having explicit errors would make it easier for us to detect bugs. The code is illustrated in the following snippet:

```
export const useCount1 = () => {
  const value = useContext(Count1Context);
```

```
  if (value === null) throw new Error("Provider missing");
  return value;
};
```

Following on, we create `Count2Context`, define a `Count2Provider` component and a `useCount2` hook (they are the same as `Count1Context`, `Count1Provider`, and `useCount1`, except for the names). The code is illustrated in the following snippet:

```
const Count2Context = createContext<
  CountContextType | null
>(null);

export const Count2Provider = ({
  children
}: {
  children: ReactNode
}) => (
  <Count2Context.Provider value={useState(0)}>
    {children}
  </Count2Context.Provider>
);

export const useCount2 = () => {
  const value = useContext(Count2Context);
  if (value === null) throw new Error("Provider missing");
  return value;
};
```

Next, we define a `Counter1` component to use the `count1` state and show the count and a button. Notice in the following code snippet that this component doesn't know about a Context, which is hidden in the `useCount1` hook:

```
const Counter1 = () => {
  const [count1, setCount1] = useCount1();
  return (
    <div>
      Count1: {count1}
      <button onClick={() => setCount1((c) => c + 1)}>
        +1
```

```
      </button>
    </div>
  );
};
```

Likewise, we define a `Counter2` component, as follows:

```
const Counter2 = () => {
  const [count2, setCount2] = useCount2();
  return (
    <div>
      Count2: {count2}
      <button onClick={() => setCount2((c) => c + 1)}>
        +1
      </button>
    </div>
  );
};
```

Notice the `Counter2` component is almost similar to the `Counter1` component. The major difference is that the `Counter2` component uses the `useCount2` hook instead of the `useCount1` hook.

We define a `Parent` component that has `Counter1` and `Counter2` defined previously, as follows:

```
const Parent = () => (
  <div>
    <Counter1 />
    <Counter1 />
    <Counter2 />
    <Counter2 />
  </div>
);
```

Finally, an App component is defined to complete the example. It wraps the `Parent` component with two provider components, as follows:

```
const App = () => (
  <Count1Provider>
    <Count2Provider>
      <Parent />
    </Count2Provider>
  </Count1Provider>
);
```

Although it's not very explicit with this snippet, we typically have a separate file such as `contexts/count1.jsx` for each Context and export only custom hooks such as `useCount1` and provider components such as `Count1Provider`. In this case, `Count1Context` is not exported.

Factory pattern with a custom hook

Creating a custom hook and a provider component is a somewhat repetitive task; however, we can create a function that does the task.

The following example shows `createStateContext` as a concrete implementation.

The `createStateContext` function takes a `useValue` custom hook that takes an initial value and returns a state. If you use `useState`, it returns a tuple of the `state` value and the `setState` function. The `createStateContext` function returns a tuple of a provider component and a custom hook to get the state. This is the pattern we learned in the previous sections.

In addition, this provides a new feature; the provider component accepts an optional `initialValue` prop that is passed into `useValue`. This allows you to set the initial value of the state at runtime instead of defining an initial value at creation. The code is illustrated in the following snippet:

```
const createStateContext = (
  useValue: (init) => State,
) => {
  const StateContext = createContext(null);
  const StateProvider = ({
    initialValue,
    children,
```

```
  }) => (
    <StateContext.Provider value={useValue(initialValue)}>
      {children}
    </StateContext.Provider>
  );
  const useContextState = () => {
    const value = useContext(StateContext);
    if (value === null) throw new Error("Provider
      missing");
    return value;
  };
  return [StateProvider, useContextState] as const;
};
```

Now, let's see how to use createStateContext. We define a custom hook,
useNumberState; it takes an optional init parameter. We then invoke useState
with init, as follows:

```
const useNumberState = (init) => useState(init || 0);
```

By passing useNumberState to createStateContext, we can create as many
state Contexts as we want; we created two sets of them. The types of useCount1
and useCount2 are inferred from useNumberState. The code is illustrated in
the following snippet:

```
const [Count1Provider, useCount1] =
  createStateContext(useNumberState);
const [Count2Provider, useCount2] =
  createStateContext(useNumberState);
```

Notice we avoid the repetitive definition thanks to createStateContext.

We then define Counter1 and Counter2 components. The way to use useCount1
and useCount2 is identical to the previous example, as we can see in the following
code snippet:

```
const Counter1 = () => {
  const [count1, setCount1] = useCount1();
  return (
    <div>
```

```
          Count1: {count1}
          <button onClick={() => setCount1((c) => c + 1)}>
            +1
          </button>
        </div>
    );
};

const Counter2 = () => {
    const [count2, setCount2] = useCount2();
    return (
        <div>
          Count2: {count2}
          <button onClick={() => setCount2((c) => c + 1)}>
            +1
          </button>
        </div>
    );
};
```

Finally, we create `Parent` and `App` components. The way to use `Count1Provider` and `Count2Provider` is also the same, as we can see here:

```
const Parent = () => (
    <div>
      <Counter1 />
      <Counter1 />
      <Counter2 />
      <Counter2 />
    </div>
);

const App = () => (
    <Count1Provider>
      <Count2Provider>
        <Parent />
      </Count2Provider>
```

```
      </Count1Provider>
  );
```

Notice how we reduce our code from the previous example. The whole point of createStateContext is to avoid repetitive code and provide the same functionality.

Instead of useNumberState with useState, we could make the custom hook with useReducer, as follows:

```
const useMyState = () => useReducer({}, (prev, action) => {
  if (action.type === 'SET_FOO') {
    return { ...prev, foo: action.foo };
  }
  // ...
};
```

We could also create a more complex hook. The following example has inc1 and inc2 custom action functions. It uses useEffect to show an updated log in the console:

```
const useMyState = (initialState = { count1: 0, count2: 0 }) =>
{
  const [state, setState] = useState(initialState);
  useEffect(() => {
    console.log('updated', state);
  });
  const inc1 = useCallback(() => {
    setState((prev) => ({
      ...prev,
      count1: prev.count1 + 1
    }));
  }, []);
  const inc2 = useCallback(() => {
    setState((prev) => ({
      ...prev,
      count2: prev.count2 + 1
    }));
  }, []);
  return [state, { inc1, inc2 }];
};
```

We can still use the `createStateContext` function for these `useMyState` hooks and any other custom hooks.

It's worth noting that this factory pattern works well in TypeScript. TypeScript provides extra checks with types, and developers can get better experience from type checking. The following code snippet shows the typed version of `createStateContext` and `useNumberState`:

```
const createStateContext = <Value, State>(
  useValue: (init?: Value) => State
) => {
  const StateContext = createContext<State | null>(null);
  const StateProvider = ({
    initialValue,
    children,
  }: {
    initialValue?: Value;
    children?: ReactNode;
  }) => (
    <StateContext.Provider value={useValue(initialValue)}>
      {children}
    </StateContext.Provider>
  );
  const useContextState = () => {
    const value = useContext(StateContext);
    if (value === null){
      throw new Error("Provider missing");
    }
    return value;
  };
  return [StateProvider, useContextState] as const;
};


const useNumberState = (init?: number) => useState(init || 0);
```

If we use the typed version of `createStateContext` and `useNumberState`, the result is also typed.

Avoiding provider nesting with reduceRight

With the `createStateContext` function, it's very easy to create many states.
Let's suppose we created five of them, as follows:

```
const [Count1Provider, useCount1] =
  createStateContext(useNumberState);
const [Count2Provider, useCount2] =
  createStateContext(useNumberState);
const [Count3Provider, useCount3] =
  createStateContext(useNumberState);
const [Count4Provider, useCount4] =
  createStateContext(useNumberState);
const [Count5Provider, useCount5] =
  createStateContext(useNumberState);
```

Our App component would then look like this:

```
const App = () => (
  <Count1Provider initialValue={10}>
    <Count2Provider initialValue={20}>
      <Count3Provider initialValue={30}>
        <Count4Provider initialValue={40}>
          <Count5Provider initialValue={50}>
            <Parent />
          </Count5Provider>
        </Count4Provider>
      </Count3Provider>
    </Count2Provider>
  </Count1Provider>
);
```

This is absolutely correct, and it captures how a component tree is structured. However, too much nesting is not very comfortable while coding. To mitigate this coding style, we could use `reduceRight`. The App component can be refactored, as shown in the following example:

```
const App = () => {
  const providers = [
    [Count1Provider, { initialValue: 10 }],
```

```
    [Count2Provider, { initialValue: 20 }],
    [Count3Provider, { initialValue: 30 }],
    [Count4Provider, { initialValue: 40 }],
    [Count5Provider, { initialValue: 50 }],
  ] as const;
  return providers.reduceRight(
    (children, [Component, props]) =>
      createElement(Component, props, children),
    <Parent />,
  );
};
```

There could be some variations of this technique, such as creating a **higher-order component (HOC)**, but the key point is using `reduceRight` to construct a provider tree.

This technique is not only for a global state with Context but also for any components.

In this section, we learned some best practices to work with a global state with Contexts. These are not something you must follow. As long as you understand the behavior of Context and its limitations, any pattern would work fine.

Summary

In this chapter, we learned how to create global states with React Context. The Context propagation works to avoid passing props. If you understand the Context behavior correctly, implementing global states with Context is straightforward. Basically, we should create a Context for each state piece to avoid extra re-renders. Some best practices will help in the implementation of a global state with Context, particularly the concrete implementation of `createStateContext`, which will help when organizing your app code.

In the next chapter, we will learn another pattern of implementing a global state with subscriptions.

4
Sharing Module State with Subscription

In the previous chapter, we learned how to use Context for a global state. As discussed, Context is not designed for the singleton pattern; it's a mechanism for avoiding the singleton pattern and providing different values for different subtrees. For a singleton-like a global state, using a module state makes more sense because it's a singleton value in memory. The goal of this chapter is to learn to use a module state with React. It's a less well-known pattern than Context, but is often used to integrate the existing module state.

> **What Is a Module State?**
>
> A strict definition of a module state would be some constants or variables defined in **ECMAScript (ES)** module scopes. In this book, we aren't following the strict definition. You can simply assume that a module state is a variable defined globally or within the scope of a file.

We'll explore how to use a module state as a global state in React. In order to use a module state in React components, we use a subscription mechanism.

In this chapter, we will cover the following topics:

- Exploring the module state
- Using a module state as a global state in React
- Adding a basic Subscription
- Working with a selector and `useSubscription`

Technical requirements

You are expected to have a moderate knowledge of React, including React Hooks. Refer to the official site at `https://reactjs.org` to learn more.

In some code, we use TypeScript (`https://www.typescriptlang.org`) and you should have a basic knowledge of this.

The code in this chapter is available on GitHub:

`https://github.com/PacktPublishing/Micro-State-Management-with-React-Hooks/tree/main/chapter_04`

To run code snippets, you need a React environment, for example, Create React App (`https://create-react-app.dev`) or CodeSandbox (`https://codesandbox.io`).

Exploring the module state

The module state is a variable defined at the module level. *Module* here means an ES module or just a file. For simplicity, we assume that a variable defined outside a function is a module state.

For example, let's define the `count` state:

```
let count = 0;
```

Assuming this is defined in a module, this is a module state.

Typically, with React, we want to have an object state. The following defines an object state with `count`:

```
let state = {
  count: 0,
};
```

More properties can be added to the object. Nesting objects are also possible.

Now, let's define functions to access this module state. `getState` is a function to read `state`, and `setState` is a function to write `state`:

```
export const getState = () => state;

export const setState = (nextState) => {
  state = nextState;
};
```

Notice that we added `export` to these functions to express that they are expected to be used outside the module.

In React, we often update a state with functions. Let's modify `setState` to allow a `function` update:

```
export const setState = (nextState) => {
  state = typeof nextState === 'function'
    ? nextState(state) : nextState;
};
```

You can use a function update as follows:

```
setState((prevState) => ({
  ...prevState,
  count: prevState.count + 1
}));
```

Instead of defining a module state directly, we can create a function for creating a container that includes `state` and some access functions.

The following is the concrete implementation of such a function:

```
export const createContainer = (initialState) => {
  let state = initialState;
  const getState = () => state;
  const setState = (nextState) => {
    state = typeof nextState === 'function'
      ? nextState(state) : nextState;
  };
```

```
    return { getState, setState };
};
```

You can use this as follows:

```
import { createContainer } from '...';

const { getState, setState } = createContainer({
    count: 0
});
```

So far, a module state has nothing to do with React. In the next section, we'll learn how to use a module state with React.

Using a module state as a global state in React

As we discussed in *Chapter 3, Sharing Component State with Context*, React Context is designed to provide different values for different subtrees. Using React Context for a singleton global state is a valid operation, but it doesn't use the full capability of Context.

If what you need is a global state for an entire tree, a module state might fit better. However, to use a module state in a React component, we need to handle re-rendering ourselves.

Let's start with a simple example. Unfortunately, this is a non-working example:

```
let count = 0;

const Component1 = () => {
    const inc = () => {
        count += 1;
    }
    return (
        <div>{count} <button onClick={inc}>+1</button></div>
    );
};
```

You will see count 0 at the beginning. Clicking button increases the count variable, but it doesn't trigger the component to re-render.

At the time of writing this book, React has only two hooks, `useState` and `useReducer`, to trigger re-renders. We need to use either of those to make a component reactive with a module state.

The previous example can work with the following modification:

```
let count = 0;

const Component1 = () => {
  const [state, setState] = useState(count);
  const inc = () => {
    count += 1;
    setState(count);
  }
  return (
    <div>{state} <button onClick={inc}>+1</button></div>
  );
};
```

Now, if you click `button`, it will increase the `count` variable, as well as trigger the component.

Let's see what happens if we have another component like the following:

```
const Component2 = () => {
  const [state, setState] = useState(count);
  const inc2 = () => {
    count += 2;
    setState(count);
  }
  return (
    <div>{state} <button onClick={inc2}>+2</button></div>
  );
};
```

Even if you click `button` in `Component1`, it won't trigger `Component2` to re-render. Only when you click `button` in `Component2` will it re-render and show the latest module state. This is the inconsistency between `Component1` and `Component2`, and our expectation is that both components should show the same value. The inconsistency also happens with two `Component1` components.

One naive approach to this problem is to invoke `setState` functions in `Component1` and `Component2` at the same time. This requires having `setState` functions at the module level. We should also consider the component life cycle and use the `useEffect` hook to modify a set that holds `setState` functions outside React.

The following example is one possible solution. This is to illustrate the idea and is not very practical:

```
let count = 0;
const setStateFunctions =
  new Set<(count: number) => void>();

const Component1 = () => {
  const [state, setState] = useState(count);
  useEffect(() => {
    setStateFunctions.add(setState);
    return () => { setStateFunctions.delete(setState); };
  }, []);
  const inc = () => {
    count += 1;
    setStateFunctions.forEach((fn) => {
      fn(count);
    });
  }
  return (
    <div>{state} <button onClick={inc}>+1</button></div>
  );
};
```

Notice we return a function in `useEffect` to clean up the effect. In the `inc` function, we invoke all `setState` functions in the `setStateFunctions` set.

Now, `Component2` will also be modified like `Component1`:

```
const Component2 = () => {
  const [state, setState] = useState(count);
  useEffect(() => {
    setStateFunctions.add(setState);
    return () => { setStateFunctions.delete(setState); };
```

```
  }, []);
  const inc2 = () => {
    count += 2;
    setStateFunctions.forEach((fn) => {
      fn(count);
    });
  }
  return (
    <div>{state} <button onClick={inc2}>+2</button></div>
  );
};
```

As noted, this is not a very practical solution. We have some repetitive code in `Component1` and `Component2`.

In the next section, we will introduce a Subscription mechanism and reduce the repetitive code.

Adding a basic Subscription

Here, we'll learn about the Subscription mechanism and how to connect a module state to the React state.

Subscription is a way to get notified of things such as updates. A typical use of a Subscription would look like the following:

```
const unsubscribe = store.subscribe(() => {
  console.log('store is updated');
});
```

Here, we assume a `store` variable to have a `subscribe` method that takes a `callback` function and returns an `unsubscribe` function.

In this case, the expected behavior is that whenever `store` is updated, the callback function is invoked and it shows the console log.

Now, let's implement a module state with a Subscription. We'll call it `store`, which holds the `state` value and the `subscribe` method, in addition to the `getState` and `setState` methods that we described in the *Exploring the module state* section. A `createStore` is a function to create `store` with an initial state value:

```
type Store<T> = {
  getState: () => T;
  setState: (action: T | ((prev: T) => T)) => void;
  subscribe: (callback: () => void) => () => void;
};

const createStore = <T extends unknown>(
  initialState: T
): Store<T> => {
  let state = initialState;
  const callbacks = new Set<() => void>();
  const getState = () => state;
  const setState = (nextState: T | ((prev: T) => T)) => {
    state =
      typeof nextState === "function"
        ? (nextState as (prev: T) => T)(state)
        : nextState;
    callbacks.forEach((callback) => callback());
  };
  const subscribe = (callback: () => void) => {
    callbacks.add(callback);
    return () => {
      callbacks.delete(callback);
    };
  };
  return { getState, setState, subscribe };
};
```

Compared with the `createContainer` function that we implemented in the *Exploring the module state* section, `createStore` has the `subscribe` method and the `setState` method, which invokes callbacks.

We use `createStore` as follows:

```
import { createStore } from '...';

const store = createStore({ count: 0 });
console.log(store.getState());
store.setState({ count: 1 });
store.subscribe(...);
```

The `store` variable holds `state` in it, and the entire `store` variable can be seen as a module state.

Next up is the use of the `store` variable in React.

We define a new hook, `useStore`, which will return a tuple of the `store` state value and its update function:

```
const useStore = (store) => {
  const [state, setState] = useState(store.getState());
  useEffect(() => {
    const unsubscribe = store.subscribe(() => {
      setState(store.getState());
    });
    setState(store.getState()); // [1]
    return unsubscribe;
  }, [store]);
  return [state, store.setState];
};
```

You may notice **[1]**. This is to cover an edge case. It invokes the `setState()` function once in `useEffect`. This is due to the fact that `useEffect` is delayed and there's a chance that `store` already has a new state.

The following is a component with `useStore`:

```
const Component1 = () => {
  const [state, setState] = useStore(store);
  const inc = () => {
    setState((prev) => ({
      ...prev,
      count: prev.count + 1,
```

```
      }));
  };
  return (
    <div>
      {state.count} <button onClick={inc}>+1</button>
    </div>
  );
};
```

It's important to update a module state immutably, the same as the React state, because a module state is eventually set in the React state:

Like `Component1`, we define another one, `Component2`, as follows:

```
const Component2 = () => {
  const [state, setState] = useStore(store);
  const inc2 = () => {
    setState((prev) => ({
      ...prev,
      count: prev.count + 2,
    }));
  };
  return (
    <div>
      {state.count} <button onClick={inc2}>+2</button>
    </div>
  );
};
```

Both buttons in the two components will update the module state in `store` and the states in both components are shared.

Finally, we define the `App` component:

```
const App = () => (
  <>
    <Component1 />
    <Component2 />
  </>
);
```

When you run this app, you will see something like *Figure 4.1*. If you click either the **+1** or **+2** buttons, you will see that both counts (shown as **3**) are updated together:

$$3 \quad \boxed{+1}$$
$$3 \quad \boxed{+2}$$

Figure 4.1 – Screenshot of the running app

In this section, we used a Subscription to connect the module state to a React component.

In the next section, we will use a selector function to use only part of the state, as well as learn how to use `useSubscription`.

Working with a selector and useSubscription

The `useStore` hook we created in the previous section returns a whole state object. This means that any small part of the state object change will notify all `useStore` hooks and it can cause extra re-renders.

To avoid extra re-renders, we can introduce a selector to return the only part of the state that a component is interested in.

Let's first develop `useStoreSelector`.

We use the same `createStore` function defined in the previous section and create a `store` variable as follows:

```
const store = createStore({ count1: 0, count2: 0 });
```

The state in `store` has two counts – `count1` and `count2`.

The `useStoreSelector` hook is similar to `useStore`, but it receives an additional selector function. It uses the selector function to scope the state:

```
const useStoreSelector = <T, S>(
  store: Store<T>,
  selector: (state: T) => S
) => {
  const [state, setState] =
    useState(() => selector(store.getState()));
  useEffect(() => {
    const unsubscribe = store.subscribe(() => {
      setState(selector(store.getState()));
```

```
      });
      setState(selector(store.getState()));
      return unsubscribe;
    }, [store, selector]);
    return state;
};
```

Compared to `useStore`, the `useState` hook in `useStoreSelector` holds the return value of `selector` instead of the entire state.

Now we define a component to use `useStoreSelector`. The return value of `useStoreSelector` is a count number. To update the state, we invoke `store.setState()` directly in this case. `Component1` is a component for displaying `count1` in the state:

```
const Component1 = () => {
  const state = useStoreSelector(
    store,
    useCallback((state) => state.count1, []),
  );
  const inc = () => {
    store.setState((prev) => ({
      ...prev,
      count1: prev.count1 + 1,
    }));
  };
  return (
    <div>
      count1: {state} <button onClick={inc}>+1</button>
    </div>
  );
};
```

Notice we need to use `useCallback` to get a stable selector function. Otherwise, as the selector is specified in the second argument of `useEffect`, `Component1` will subscribe to the `store` variable every time `Component1` renders.

We define `Component2`, which is to display `count2` instead of `count1`. We define a selector function outside the component to avoid `useCallback` this time:

```
const selectCount2 = (
  state: ReturnType<typeof store.getState>
) => state.count2;

const Component2 = () => {
  const state = useStoreSelector(store, selectCount2);
  const inc = () => {
    store.setState((prev) => ({
      ...prev,
      count2: prev.count2 + 1,
    }));
  };
  return (
    <div>
      count2: {state} <button onClick={inc}>+1</button>
    </div>
  );
};
```

Finally, the App component renders two components for each `Component1` component and `Component2` component for demonstration:

```
const App = () => (
  <>
    <Component1 />
    <Component1 />
    <Component2 />
    <Component2 />
  </>
);
```

Figure 4.2 is a screenshot of the running app:

count1: 5   +1
count1: 5   +1
count2: 3   +1
count2: 3   +1

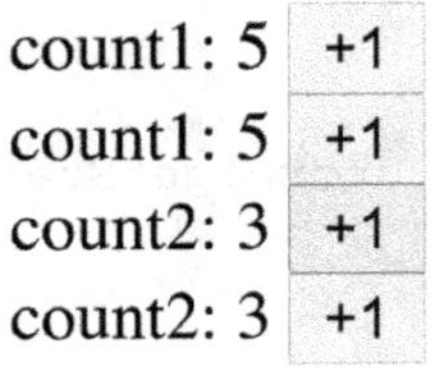

Figure 4.2 – Screenshot of the running app

The first two lines in the preceding figure are rendered by `Component1`. If you click either of the first two **+1** buttons, it will increment `count1`, which will trigger `Component1` to re-render. However, `Component2` (the last two lines in *Figure 4.2*) won't re-render because `count2` isn't changed.

While the `useStoreSelector` hook works well and is usable in production, there's a caveat when `store` or `selector` is changed. Because `useEffect` fires a little later, it will return a stale state value until re-subscribing is done. We could fix it by ourselves, but it's a little technical.

Fortunately, the React team provides an official hook for this use case. It's called `use-subscription` (`https://www.npmjs.com/package/use-subscription`).

Let's re-define `useStoreSelector` using `useSubscription`. The code is as simple as the following:

```
const useStoreSelector = (store, selector) => useSubscription(
  useMemo(() => ({
    getCurrentValue: () => selector(store.getState()),
    subscribe: store.subscribe,
  }), [store, selector])
);
```

The app still works with this change.

We could avoid using the `useStoreSelector` hook and use `useSubscription` directly in `Component1`:

```
const Component1 = () => {
  const state = useSubscription(useMemo(() => ({
    getCurrentValue: () => store.getState().count1,
    subscribe: store.subscribe,
```

```
  }), []));
  const inc = () => {
    store.setState((prev) => ({
      ...prev,
      count1: prev.count1 + 1,
    }));
  };
  return (
    <div>
      count1: {state} <button onClick={inc}>+1</button>
    </div>
  );
};
```

In this case, as useMemo is already used, useCallback is not necessary.

---

useSubscription and useSyncExternalStore

In future versions of React, a hook called useSyncExternalStore
will be included. This is a successor of useSubscription. Hence, using
the module state will become more accessible (https://github.com/
reactwg/react-18/discussions/86).

In this section, we learned about using selectors to scope state and also the official
useSubscription hook to have a more concrete solution.

Summary

In this chapter, we learned how to create a module state and integrate it in React. Using
what we learned, you can use the module state as a global state in React. Subscription
plays an important role in integration because it allows the re-rendering of components
to be triggered when the module state is changed. In addition to the basic Subscription
implementation to use the module state in React, there is an official package. Both the
basic Subscription and the official package work for the production use case.

In the next chapter, we will learn about the third pattern of implementing a global state,
which is a combination of the first pattern and the second pattern.

5
Sharing Component State with Context and Subscription

In the previous two chapters, we learned how to use Context and Subscription for a global state. Each has different benefits: Context allows us to provide different values for different subtrees, while Subscriptions prevent extra re-renders.

In this chapter, we will learn a new approach: combining React Context and Subscriptions. The combination will give us the benefits of each, which means:

- Context can provide a global state to a subtree and the Context provider can be nested. Context allows us to control a global state in the React component lifecycle like the useState hook.

- On the other hand, Subscriptions allow us to control re-renders, which is not possible with a single Context.

Having the benefits of both can be a good solution for larger apps – because, as mentioned, this means we can have different values in different subtrees, and we can also avoid extra re-renders.

This approach is useful for mid to large apps. In such apps, having different values in different subtrees can happen, and we can avoid extra re-renders, which can be very important for our apps.

In this chapter, we will cover the following topics:

- Exploring the limitations of module state
- Understanding when to use Context
- Implementing the Context and Subscription pattern

Technical requirements

You are expected to have moderate knowledge of React, including React Hooks. Refer to the official site, https://reactjs.org, to learn more.

In some code, we use TypeScript (https://www.typescriptlang.org), and you should have basic knowledge of it.

The code in this chapter is available on GitHub: https://github.com/ PacktPublishing/Micro-State-Management-with-React-Hooks/tree/ main/chapter_05.

To run code snippets, you need a React environment, for example, Create React App (https://create-react-app.dev) or CodeSandbox (https://codesandbox. io).

Exploring the limitations of module state

Because module state resides outside React components, there's a limitation: the module state defined globally is a singleton, and you can't have different states for different component trees or subtrees.

Let's revisit our `createStore` implementation from *Chapter 4, Sharing Module State with Subscription*:

```
const createStore = (initialState) => {
  let state = initialState;
  const callbacks = new Set();
  const getState = () => state;
  const setState = (nextState) => {
    state = typeof nextState === 'function'
```

```
      ? nextState(state) : nextState;
    callbacks.forEach((callback) => callback());
  };
  const subscribe =(callback) => {
    callbacks.add(callback);
    return () => { callbacks.delete(callback); };
  };
  return { getState, setState, subscribe };
};
```

Using this `createStore`, let's define a new `store`. We define a `store` with a property `count`:

```
const store = createStore({ count: 0 });
```

Note that this `store` is defined outside the React component.

To use `store` in a React component, we use `useStore`. The following is an example with two components that show the shared count from the same `store` variable. We use `useStore`, which was defined in *Chapter 4, Sharing Module State with Subscription*:

```
const Counter = () => {
  const [state, setState] = useStore(store);
  const inc = () => {
    setState((prev) => ({
      ...prev,
      count: prev.count + 1,
    }));
  };
  return (
    <div>
      {state.count} <button onClick={inc}>+1</button>
    </div>
  );
};

const Component = () => (
  <>
    <Counter />
```

```
    <Counter />
  </>
);
```

We have the component `Counter`, which is to show the `count` number in the `store` object, and a `button` to update the `count` value. Because this `Counter` component is reusable, `Component` can have two `Counter` instances. This will show a pair of two counters sharing the same state.

Now, suppose we want to show another pair of counters. We would like to have two new components in `Component`, but the new pair should show different counters from the first set.

Let's create a new `count` value. We could add a new property to the `store` object we already defined, but we assume there are other properties and want to isolate stores. Therefore, we create `store2`:

```
const store2 = createStore({ count: 0 })
```

Because `createStore` is reusable, creating a new `store2` object is straightforward.

We then need to create components to use `store2`:

```
const Counter2 = () => {
  const [state, setState] = useStore(store2);
  const inc = () => {
    setState((prev) => ({
      ...prev,
      count: prev.count + 1,
    }));
  };
  return (
    <div>
      {state.count} <button onClick={inc}>+1</button>
    </div>
  );
};

const Component2 = () => (
  <>
    <Counter2 />
    <Counter2 />
```

```
    </>
  );
```

You may notice the similarity between `Counter` and `Counter2` – that they are both 14 lines of code, and the only difference is the `store` variable they are referencing – `store` for `Counter` and `store2` for `Counter2`. We would need `Counter3` or `Counter4` to support more stores. Ideally, `Counter` should be reusable. But, as module state is defined outside React, it's not possible. This is the limitation of module state.

> **Important Note**
> You may notice we can make the `Counter` component reusable if we put `store` in `props`. However, that will require prop drilling when components are deeply nested, and the primary reason for introducing module state is to avoid prop drilling.

It would be nice to reuse the `Counter` component for different stores. The pseudocode would be as follows:

```
const Component = () => (
  <StoreProvider>
    <Counter />
    <Counter />
  </StoreProvider>
);

const Component2 = () => (
  <Store2Provider>
    <Counter />
    <Counter />
  </Store2Provider>
);

const Component3 = () => (
  <Store3Provider>
    <Counter />
    <Counter />
  </Store3Provider>
);
```

If you look at the code, you will notice that `Component`, `Component2`, and `Component3` are mostly the same. The only difference is the `Provider` components. This is exactly where React Context fits in. We will discuss this in more detail in the *Implementing the Context and Subscription pattern* section.

Now you understand the limitation of module state and the ideal patterns for multiple stores. Next up, we'll recap React Context and explore the usage of Context.

Understanding when to use Context

Before diving into learning the way to combine Context and Subscription, let's recap how Context works.

The following is a simple Context example with a theme. So, we specify a default value for `createContext`:

```
const ThemeContext = createContext("light");

const Component = () => {
  const theme = useContext(ThemeContext);
  return <div>Theme: {theme}</div>
};
```

What `useContext(ThemeContext)` returns depends on the Context in the component tree.

To change the Context value, we use a `Provider` component in Context as follows:

```
<ThemeContext.Provider value="dark">
  <Component />
</ThemeContext.Provider>
```

In this case, `Component` will show the theme as `dark`.

The provider can be nested. It will use the value from the innermost provider:

```
<ThemeContext.Provider value="this value is not used">
  <ThemeContext.Provider value="this value is not used">
    <ThemeContext.Provider value="this is the value used">
      <Component />
    </ThemeContext.Provider>
  </ThemeContext.Provider>
</ThemeContext.Provider>
```

If there are no providers in the component tree, it will use the default value.

For example, here, we assume `Root` is a component at the root:

```
const Root = () => (
  <>
    <Component />
  </>
);
```

In this case, `Component` will show the theme as `light`.

Let's see an example that has a provider to provide the same default value at the root:

```
const Root = () => (
  <ThemeContext.Provider value="light">
    <Component />
  </ThemeContext.Provider>
);
```

In this case too, `Component` will show the theme as `light`.

So, let's discuss when to use Context. To do this, think of our example: what is the difference between this example with a provider and the previous example without a provider? We can say that there is no difference. Using the default value gives the same result.

Having a proper default value for Context is important. The Context provider can be seen as a method to override the default Context value or a value provided by the parent provider if it exists.

In the case of `ThemeContext`, if we have the proper default value, then what's the point of using a provider? It will be required to provide a different value for a subtree of the entire component tree. Otherwise, we can just use the default value from `Context`.

For a global state with Context, you may only use one provider at the root. This is a valid use case, but this use case can be covered by module state with Subscription, which we learned about in *Chapter 4, Sharing Module State with Subscription*. Given that module state covers the use case with one Context provider at the root, Context for a global state is only required if we need to provide different values for different subtrees.

In this section, we revisited React Context and learned when to use it. Next up, we will learn how to combine Context and Subscription.

Implementing the Context and Subscription pattern

As we learned, using one Context to propagate a global state value has a limitation: it causes extra re-renders.

Module state with Subscription doesn't have such a limitation, but there is another: it only provides a single value for the entire component tree.

We would like to combine Context and Subscription to overcome both limitations. Let's implement this feature. We'll start with `createStore`. This is exactly the same implementation we developed in *Chapter 4, Sharing Module State with Subscription*:

```
type Store<T> = {
  getState: () => T;
  setState: (action: T | ((prev: T) => T)) => void;
  subscribe: (callback: () => void) => () => void;
};


const createStore = <T extends unknown>(
  initialState: T
): Store<T> => {
  let state = initialState;
  const callbacks = new Set<() => void>();
  const getState = () => state;
  const setState = (nextState: T | ((prev: T) => T)) => {
    state =
      typeof nextState === "function"
        ? (nextState as (prev: T) => T)(state)
        : nextState;
    callbacks.forEach((callback) => callback());
  };
  const subscribe = (callback: () => void) => {
    callbacks.add(callback);
    return () => {
      callbacks.delete(callback);
    };
  };
```

```
  return { getState, setState, subscribe };
};
```

In *Chapter 4, Sharing Module State with Subscription*, we used `createStore` for module state. This time, we'll use `createStore` for the `Context` value.

The following is the code to create a Context. The default value is passed to `createContext`, which we refer to as a default store:

```
type State = { count: number; text?: string };

const StoreContext = createContext<Store<State>>(
  createStore<State>({ count: 0, text: "hello" })
);
```

In this case, the default store has a state with two properties: `count` and `text`.

To provide different stores for subtrees, we implement `StoreProvider`, which is a tiny wrapper around `StoreContext.Provider`:

```
const StoreProvider = ({
  initialState,
  children,
}: {
  initialState: State;
  children: ReactNode;
}) => {
  const storeRef = useRef<Store<State>>();
  if (!storeRef.current) {
    storeRef.current = createStore(initialState);
  }
  return (
    <StoreContext.Provider value={storeRef.current}>
      {children}
    </StoreContext.Provider>
  );
};
```

`useRef` is used to make sure that the store object is initialized only once at the first render.

To use a store object, we implement a hook called `useSelector`. Unlike `useStoreSelector`, defined in the *Working with a selector and useSubscription* section in *Chapter 4, Sharing Module State with Subscription*, `useSelector` doesn't take a `store` object in its arguments. It takes a `store` object from `StoreContext` instead:

```
const useSelector = <S extends unknown>(
  selector: (state: State) => S
) => {
  const store = useContext(StoreContext);
  return useSubscription(
    useMemo(
      () => ({
        getCurrentValue: () => selector(store.getState()),
        subscribe: store.subscribe,
      }),
      [store, selector]
    )
  );
};
```

Using `useContext` together with `useSubscription` is the key point of this pattern. This combination allows us the benefits of both Context and Subscription.

Unlike module state, we need to provide a way to update the state with Context. `useSetState` is a simple hook to return the `setState` function in `store`:

```
const useSetState = () => {
  const store = useContext(StoreContext);
  return store.setState;
};
```

Now, let's use what we have implemented. The following is a component that shows `count` in `store`, along with `button` to increment `count`. We define `selectCount` outside the `Component`, otherwise, we would need to wrap the function with `useCallback`, which introduces extra work:

```
const selectCount = (state: State) => state.count;

const Component = () => {
  const count = useSelector(selectCount);
```

```
  const setState = useSetState();
  const inc = () => {
    setState((prev) => ({
      ...prev,
      count: prev.count + 1,
    }));
  };
  return (
    <div>
      count: {count} <button onClick={inc}>+1</button>
    </div>
  );
};
```

It's important to note here that this `Component` component is not tied to any specific store object. The `Component` component can be used for different stores.

We can also have `Component` in various places:

- Outside any providers
- Inside the first provider
- Inside the second provider

The following `App` component includes the `Component` components in three places: 1) outside of `StoreProvider`, 2) inside the first `StoreProvider` component, and 3) inside the second nested `StoreProvider` component. The `Component` components in different `StoreProvider` components share different `count` values:

```
const App = () => (
  <>
    <h1>Using default store</h1>
    <Component />
    <Component />
    <StoreProvider initialState={{ count: 10 }}>
      <h1>Using store provider</h1>
      <Component />
      <Component />
      <StoreProvider initialState={{ count: 20 }}>
        <h1>Using inner store provider</h1>
```

```
        <Component />
        <Component />
      </StoreProvider>
    </StoreProvider>
  </>
);
```

Each `Component` component using the same `store` object will share the `store` object and show the same `count` value. In this case, the components in different component tree levels use a different `store`, hence the components show a different `count` value in various places. When you run this app, you will see something like the following:

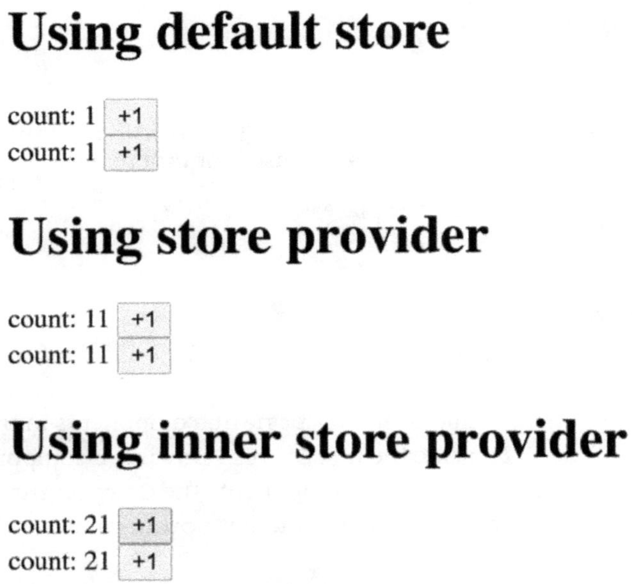

Using default store

count: 1 +1
count: 1 +1

Using store provider

count: 11 +1
count: 11 +1

Using inner store provider

count: 21 +1
count: 21 +1

Figure 5.1 – Screenshot of the running app

If you click the **+1** button in **Using default store**, you will see two counts in **Using default store** are updated together. If you click the **+1** button in **Using store provider**, you will see two counts in **Using store provider** are updated together. The same applies to **Using inner store provider**.

In this section, we learned how to implement a global state with Context and Subscription, taking advantage of the related benefits. We can isolate state in a subtree thanks to Context, and we can avoid extra re-renders thanks to Subscription.

Summary

In this chapter, we learned a new approach: combining React Context and Subscription. It provides the benefits of both: providing isolated values in subtrees and avoiding extra re-renders. This approach is useful for mid to large apps. In such apps, having different values in different subtrees can happen, and we can avoid extra re-renders, which can be very important for our apps.

Starting from the next chapter, we will dive into various global state libraries. We will learn how those libraries are based on what we have learned so far.

Part 3: Library Implementations and Their Uses

In this part, we introduce four libraries for micro-state management. We discuss their approaches for optimizing re-renders along with their use. We explain the similarities and differences among all four libraries. Finally, you will learn how to choose libraries based on their requirements and preferences.

This part comprises the following chapters:

- *Chapter 6, Introducing Global State Libraries*
- *Chapter 7, Use Case Scenario 1 – Zustand*
- *Chapter 8, Use Case Scenario 2 – Jotai*
- *Chapter 9, Use Case Scenario 3 – Valtio*
- *Chapter 10, Use Case Scenario 4 – React Tracked*
- *Chapter 11, Similarities and Differences between Three Global State Libraries*

6
Introducing Global State Libraries

We have learned about several patterns used to share state among components so far. The rest of this book will introduce various global state libraries that use such patterns.

Before diving into the libraries, we will recap the challenges associated with global states and discuss two aspects of libraries: where the state resides and how to control re-renders. With this in hand, we will be able to understand the characteristics of global state libraries.

In this chapter, we will cover the following topics:

- Working with global state management issues
- Using the data-centric and component-centric approaches
- Optimizing re-renders

Technical requirements

You are expected to have moderate knowledge of React, including React hooks. Refer to the official site at `https://reactjs.org` to learn more.

To run the code snippets, you need a React environment, for example, Create React App (`https://create-react-app.dev`) or CodeSandbox (`https://codesandbox.io`).

Working with global state management issues

React is designed around the concept of components. In the component model, everything is expected to be reusable. Global state is something that exists outside of components. It's often true that we should avoid using a global state where possible because it requires an extra dependency on a component. However, a global state is sometimes very handy and allows us to be more productive. For some app requirements, global state fits well.

There are two challenges when designing a global state:

- The first challenge is how to read a global state.

  Global state tends to have multiple values. It's often the case that a component using a global state doesn't need all the values in it. If a component re-renders when a global state is changed but the changed values are not relevant to the component, it's an extra re-render. Extra re-renders are not desirable, and global state libraries should provide a solution for them. There are several approaches to avoiding extra re-renders, and we will discuss them in more detail in the *Optimizing re-renders* section.

- The second challenge is how to write or update a global state.

  Again, global state is likely to have multiple values, some of which may be nested objects. It might not be a good idea to have a single global variable and accept arbitrary mutations. The following code block shows an example of a global variable and one arbitrary mutation:

```
let globalVariable = {
  a: 1,
  b: {
    c: 2,
    d: 3,
  },
```

```
    e: [4, 5, 6],
  };
```

```
  globalVariable.b.d = 9;
```

The mutation `globalVariable.b.d = 9` in the example may not work for a global state because there's no way to detect the change and trigger React components to re-render.

To have more control over how to write a global state, we often provide functions to update a global state. It's also often necessary to hide a variable in a closure so that the variable can't be mutated directly. The following code block shows an example of creating two functions for reading and writing a variable in a closure:

```
const createContainer = () => {
  let state = { a: 1, b: 2 };
  const geState = () => state;
  const setState = (...) => { ... };
  return { getState, setState };
};
```

```
const globalContainer = createContainer();
globalContainer.setState(...);
```

The `createContainer` function creates `globalContainer`, which holds `getState` and `setState` functions. `getState` is a function to read a global state and `setState` is a function to update a global state. There are several ways to implement functions such as `setState` to update a global state. We will look at concrete examples in the following chapters.

Global versus General State Management

This book focuses on *global* state management; *general* state management is out of scope. In the field of general state management, popular approaches include the one-way data flow approach, as in Redux (`https://redux.js.org`), and the state machine-based approach, as in XState (`https://xstate.js.org`). General state management approaches are useful not only for a global state but also for a local state.

> **Notes about Redux and React Redux**
>
> Redux has been a big player in a global state management. Redux solves state management with one-way data flow with a global state in mind. However, Redux itself has nothing to do with React. It's React Redux (`https://react-redux.js.org`) that binds React and Redux. While Redux itself doesn't have a capability or a notion to avoid extra re-renders, React Redux has such a capability.
>
> Because Redux and React Redux were so popular, some people overused them in the past. This was due to the lack of React Context before React 16.3, and there were no other popular options. Such people (mis-)used React Redux mainly for (legacy) Context, without needing the one-way data flow. With React Context since React 16.3 and the `useContext` hook since React 16.8, we can easily solve use cases to avoid prop drilling and extra re-renders. That brings us to microstate management – our focus in this book.
>
> Hence, technically speaking, React Redux minus Redux is within the scope of this book. Redux itself is a great solution for general state management, and along with React Redux, it solves the global state issues we discussed in this section.

In this section, we discussed the general challenges when it comes to global state libraries. Next up, we will learn about where state resides.

Using the data-centric and component-centric approaches

Global state can technically be divided into two types: data-centric and component-centric.

In the following sections, we will discuss both these approaches in detail. Then, we will also talk about some exceptions.

Understanding the data-centric approach

When you design an app, you may have a data model as a singleton in your app and you may already have the data to deal with. In this case, you would define components and connect the data and the components. The data can be changed from the outside, such as by other libraries or from other servers.

For the data-centric approach, module state would fit better, because module state resides in JavaScript memory outside React. Module state can exist before React starts rendering or even after all React components are unmounted.

Global state libraries using the data-centric approach would provide APIs to create module state and to connect the module state to React components. Module state is usually wrapped in a `store` object, which has methods to access and update a `state` variable.

Understanding the component-centric approach

Unlike the data-centric approach, with the component-centric approach, you can design components first. At some point, some components may need to access shared information. As we discussed in the *Effectively using local states* section in *Chapter 2, Using Local and Global States*, we can lift state and pass it down with props (a.k.a. prop drilling). If prop drilling won't work as a solution, that's when we can introduce a global state. Certainly, we can start by designing a data model first, but in the component-centric approach, the data model is fairly tied to components.

For the component-centric approach, component state, which holds a global state in the component lifecycle, fits better. This is because when all the corresponding components are unmounted, a global state is gone too. This capability allows us to have two or more global states that exist in JavaScript memory because they are in different component subtrees (or different portals).

Global state libraries using a data-centric approach provide a factory function to create functions that initialize a global state for use in React components. A factory function doesn't directly create a global state, but by using the generated functions, we let React handle a global state lifecycle.

Exploring the exceptions of both approaches

What we have described are typical use cases, and there are always some exceptions. The data-centric approach and the component-centric approach are not really two sides of the same coin. In reality, you can use one of two approaches or a hybrid of the two approaches.

Module state is often used as a singleton pattern, but you can create multiple module states for subtrees. You can even control the lifecycles of them.

Component state is often used to provide a state in a subtree, but if you put the provider component at the root of the tree and there's only one tree in JavaScript memory, it can be treated like a singleton pattern.

Component state is often implemented with the useState hook, but if we need to have a mutable variable or store, an implementation with the useRef hook is possible. The implementation might be more complicated than using useState, but it still comes under the component lifecycle.

In this section, we learned about two approaches for using a global state. Module state is mainly for use with the data-centric approach, and component state is mainly for use with the component-centric approach. Next, we will learn about several patterns to optimize re-renders.

Optimizing re-renders

Avoiding extra re-renders is a major challenge when it comes to a global state. This is a big point to consider when designing a global state library for React.

Typically, a global state has multiple properties, and they can be nested objects. See the following, for example:

```
let state = {
  a: 1,
  b: { c: 2, d: 3 },
  e: { f: 4, g: 5 },
};
```

With this state object, let's assume two components ComponentA and ComponentB, which use state.b.c and state.e.g, respectively. The following is pseudocode of the two components:

```
const ComponentA = () => {
  return <>value: {state.b.c}</>;
};
```

```
const ComponentB = () => {
  return <>value: {state.e.g}</>;
};
```

Now, let's suppose we change state as follows:

```
++state.a;
```

This changes the a property of `state`, but it doesn't change either `state.b.c` or `state.e.g`. In this case, the two components don't need to re-render.

The goal of optimizing re-renders is to specify which part of `state` is used in a component. We have several approaches to specify the part of `state`. This section describes three approaches:

- Using a selector function
- Detecting property access
- Using atoms

We will discuss each of these now.

Using a selector function

One approach is using a selector function. A selector function takes a `state` variable and returns a part of the `state` variable.

For example, let's suppose we have a `useSelector` hook that takes a selector function and returns part of `state`:

```
const Component = () => {
  const value = useSelector((state) => state.b.c);
  return <>{value}</>;
};
```

If `state.b.c` is 2, then `Component` will show 2. Now that we know that this component cares only about `state.b.c`, we can avoid extra re-renders only when `state.a` is changed.

`useSelector` will be used to compare the selector function's result every time `state` is changed. Hence, it's important that the selector function returns the referentially equal result when given the same input.

The selector function is so flexible that it can return not only a part of `state`, but also any derived value. For example, it can return a doubled value, like here:

```
const Component = () => {
  const value = useSelector((state) => state.b.c * 2);
  return <>{value}</>;
};
```

> **A Note about Selector and Memoization**
>
> If a value returned by the selector function is a primitive value such as
> a number, there are no issues. However, if the selector function returns
> a derived object value, we need to make sure to return a referentially equal
> object with the so-called memoization technique. You can read more
> about memoization at `https://en.wikipedia.org/wiki/`
> `Memoization`.

As a selector function is a means to explicitly specify which part of a component will be
used, we call this a manual optimization.

Detecting property access

Can we do render optimization automatically, without using a selector function to
explicitly specify which part of a state is to be used in a component? There is something
called state usage tracking, which is used to detect property access and use the detected
information for render optimization.

For example, let's suppose we have a `useTrackedState` hook that has the state usage
tracking capability:

```
const Component = () => {
  const trackedState = useTrackedState();
  return <p>{trackedState.b.c}</p>;
};
```

This works as `trackedState` can detect that the `.b.c` property is accessed, and
`useTrackedState` only triggers re-renders when the `.b.c` property value is changed.
This is automatic render optimization, whereas `useSelector` is manual
render optimization.

For simplicity, the previous code block example is contrived. This example can easily be
implemented with `useSelector`, the manual render optimization. Let's look at another
example using two values:

```
const Component = () => {
  const trackedState = useTrackedState();
  return (
    <>
      <p>{trackedState.b.c}</p>
      <p>{trackedState.e.g}</p>
    </>
```

```
    );
  };
```

Now, this is surprisingly difficult to implement with a single `useSelector` hook. If we were to write a selector, it would require memoization or a custom equality function, which are complicated techniques. However, if we use `useTrackedState`, it works without such complicated techniques.

The implementation of `useTrackedState` requires Proxy (`https://developer.mozilla.org/en-US/docs/Web/JavaScript/Reference/Global_Objects/Proxy`) to trap the property access to the `state` object. If this is implemented properly, it can replace most use cases of `useSelector` and can do the automatic render optimization. However, there's a subtle case where the automatic render optimization doesn't work perfectly. Let's take a closer look in the next section.

The difference between useSelector and useTrackedState

There are some use cases in which `useSelector` works better than `useTrackedState`. Because `useSelector` can create any derived values, it can derive state into simpler values.

The difference between the working of `useSelector` and `useTrackedState` can be seen with the help of a simple example. The following is an example component with `useSelector`:

```
const Component = () => {
  const isSmall = useSelector((state) => state.a < 10);
  return <>{isSmall ? 'small' : 'big'}</>;
};
```

If we were to create the same component with `useTrackedState`, it would be the following:

```
const Component = () => {
  const isSmall = useTrackedState().a < 10;
  return <>{isSmall ? 'small' : 'big'}</>;
};
```

Functionality-wise, this component with `useTrackedState` works fine, but it will trigger re-renders every time `state.a` is changed. On the contrary, with `useSelector`, it will trigger re-renders only when `isSmall` is changed, which means it's better render optimized.

Using atoms

There's another approach, which we call using atoms. An atom is a minimal unit of state used to trigger re-renders. Instead of subscribing to the whole global state and trying to avoid extra re-renders, atoms allow you to subscribe granularly.

For example, let's suppose we have a useAtom hook that only subscribes to an atom. An atom function would create such a unit (that is, atom) of a state object:

```
const globalState = {
  a: atom(1),
  b: atom(2),
  e: atom(3),
};


const Component = () => {
  const value = useAtom(globalState.a);
  return <>{value}</>;
};
```

If atoms are completely separated, it's almost equivalent to having separate global states. However, we could create a derived value with atoms. For example, say we would like to sum the globalState values. The pseudocode would be the following:

```
const sum = globalState.a + globalState.b + globalState.c;
```

To make this work, we need to track the dependency and re-evaluate the derived value when a dependency atom is updated. We will look closely at how such an API is implemented in *Chapter 8, Use Case Scenario 2 – Jotai.*

The approach using atoms can be seen as something between a manual approach and an automatic approach. While the definition of atoms and derived values is explicit (manual), the dependency tracking is automatic.

In this section, we learned about the various patterns for optimizing re-renders. It's important for a global state library to design how to optimize re-renders. It often affects the library API, and understanding how to optimize re-renders is also worthwhile for library users.

Summary

In this chapter, before diving into the actual implementation of global state libraries, we learned about some basic challenges associated with it, and some categories to differentiate global state libraries. When choosing a global state library, we can see how the library lets us read a global state and write a global state, where the library stores a global state, and how the library optimizes re-renders. These are important aspects to understand which libraries work well for certain use cases, and they should help you to choose a library that suits your needs.

In the next chapter, we will learn about the Zustand library, a library that takes a data-centric approach and optimizes re-renders with selector functions.

7
Use Case Scenario 1 – Zustand

So far, we have been exploring some basic patterns we can use to implement a global state in React. In this chapter, we will learn about a real implementation that is publicly available as a package, called Zustand.

Zustand (`https://github.com/pmndrs/zustand`) is a tiny library primarily designed to create module state for React. It's based on an immutable update model, in which state objects can't be modified but always have to be newly created. Render optimization is done manually using selectors. It has a straightforward and yet powerful `store` creator interface.

In this chapter, we will explore how module state and subscriptions are used and see what the library API looks like.

In this chapter, we will cover the following topics:

- Understanding module state and immutable state
- Adding React hooks to optimize re-renders
- Working with read state and update state
- Handling structured data
- Pros and cons of this approach and library

Technical requirements

You are expected to have moderate knowledge of React, including React hooks. Please refer to the official site, `https://reactjs.org`, to learn more.

In some of the code in this chapter, we will be using TypeScript (`https://www.typescriptlang.org`), so you should have basic knowledge of it.

The code in this chapter is available on GitHub at `https://github.com/PacktPublishing/Micro-State-Management-with-React-Hooks/tree/main/chapter_07`.

To run the code snippets in this chapter, you will need a React environment, such as Create React App (`https://create-react-app.dev`) or CodeSandbox (`https://codesandbox.io`).

At the time of writing, the current version of Zustand is v3. Future versions may provide some different APIs.

Understanding module state and immutable state

Zustand is a library that's used to create a `store` that holds a state. It's primarily designed for module state, which means you define this `store` in a module and export it. It's based on the immutable state model, in which you are not allowed to modify state object properties. Updating states must be done by creating new objects, while unmodified state objects must be reused. The benefit of the immutable state model is that you only need to check state object referential equality to know if there's any update; you don't have to check equality deeply.

The following is a minimal example that can be used to create a `count` state. It takes a `store` creator function that returns an initial state:

```
// store.ts
import create from "zustand";

export const store = create(() => ({ count: 0 }));
```

`store` exposes some functions such as `getState`, `setState`, and `subscribe`. You can use `getState` to get the state in `store` and `setState` to set the state in `store`:

```
console.log(store.getState()); // ---> { count: 0 }
store.setState({ count: 1 });
console.log(store.getState()); // ---> { count: 1 }
```

The state is immutable, and you can't mutate it like you can `++state.count`. The following example is an invalid usage that violates the state's immutability:

```
const state1 = store.getState();
state1.count = 2; // invalid
store.setState(state1);
```

`state1.count = 2` is the invalid usage, so it doesn't work as expected. With this invalid usage, the new state has the same reference as the old state, and the library can't detect the change properly.

The state must be updated with a new object such as `store.setState({ count: 2 })`. The `store.setState` function also accepts a function to update:

```
store.setState((prev) => ({ count: prev.count + 1 }));
```

This is called a function update, and it makes it easy to update the state with the previous state.

So far, we only have one `count` property in the state. The state can have multiple properties. The following example has an additional `text` property:

```
export const store = create(() => ({
  count: 0,
  text: "hello",
}));
```

Again, the state must be updated immutably, like so:

```
store.setState({
  count: 1,
  text: "hello",
});
```

However, `store.setState()` will merge the new state and the old state. Hence, you can only specify the properties you want to set:

```
console.log(store.getState());
store.setState({
  count: 2,
});
console.log(store.getState());
```

The first `console.log` statement outputs `{ count: 1, text: 'hello' }`, while the second one outputs `{ count: 2, text: 'hello' }`.

As this only changes `count`, the `text` property isn't changed. Internally, this is implemented with `Object.assign()`, as follows:

```
Object.assign({}, oldState, newState);
```

The `Object.assign` function will return a new object by merging the `oldState` and `newState` properties.

The last piece of the `store` function is `store.subscribe`. The `store.subscribe` function allows you to register a callback function, which will be invoked every time the state in `store` is updated. It works like this:

```
store.subscribe(() => {
  console.log("store state is changed");
});


store.setState({ count: 3 });
```

With the `store.setState` statement, the **store state is changed** message will be shown on the console, thanks to the subscription. `store.subscribe` is an important function for implementing React hooks.

In this section, we learned about the basics of Zustand. You might notice that this is very close to what we learned in *Chapter 4, Sharing Module State with Subscription*. Essentially, Zustand is a thin library built around the idea of an immutable state model and subscription.

In the next section, we will learn how to use `store` in React.

Using React hooks to optimize re-renders

For global states, optimizing re-renders is important because not all components use all the properties in a global state. Let's learn how Zustand addresses this.

To use `store` in React, we need a custom hook. Zustand's `create` function creates a `store` that can be used as a hook.

To follow the naming convention of React hooks, we have named the created value `useStore` instead of `store`:

```
// store.ts
import create from "zustand";

export const useStore = create(() => ({
  count: 0,
  text: "hello",
}));
```

Next, we must use the created `useStore` hook in React components. The `useStore` hook, if it's invoked, returns the entire `state` object, including all its properties. For example, let's define a component that shows the `count` value in `store`:

```
import { useStore } from "./store.ts";

const Component = () => {
  const { count, text } = useStore();
  return <div>count: {count}</div>;
};
```

This component shows the `count` value and that whenever the `store` state is changed, it will re-render. While this works fine of the time, if only the `text` value is changed and the `count` value is not changed, the component will output essentially the same **JavaScript Syntax Extension (JSX)** element and users won't see any change onscreen. Hence, this means that changing the `text` value causes extra re-renders.

When we need to avoid extra re-renders, we can specify a selector function; that is, useStore. The previous component can be rewritten with a selector function, as follows:

```
const Component = () => {
  const count = useStore((state) => state.count);
  return <div>count: {count}</div>;
};
```

By making this change, but only when the count value is changed, the component will re-render.

This selector-based extra re-render control is what we call **manual render optimization**. The way the selector works to avoid re-renders is to compare the results of what the selector function returns. You need to be careful when you're defining a selector function to return stable results to avoid re-renders.

For example, the following example doesn't work well because the selector function creates a new array with a new object in it:

```
const Component = () => {
  const [{ count }] = useStore(
    (state) => [{ count: state.count }]
  );
  return <div>count: {count}</div>;
};
```

As a result, the component will re-render, even if the count value is unchanged. This is a pitfall when we use selectors for render optimization.

In summary, the benefit of selector-based render optimization is that the behavior is fairly predictable because you explicitly write selector functions. However, the downside of selector-based render optimization is that it requires an understanding of object references.

In this section, we learned how to use a hook that's been created with Zustand, as well as how to optimize re-renders with selectors.

Next, we will learn how to use Zustand with React using a minimal example.

Working with read state and update state

While Zustand is a library that can be used in various ways, it has a pattern to read state and update state. Let's learn how to use Zustand with a small example.

Here's our small `store` with the `count1` and `count2` properties:

```
type StoreState = {
  count1: number;
  count2: number;
};

const useStore = create<StoreState>(() => ({
  count1: 0,
  count2: 0,
}));
```

This creates a new `store` with two properties called `count1` and `count2`. Notice that `StoreState` is the `type` definition in TypeScript.

Next, we must define the `Counter1` component, which shows a `count1` value. We must define the `selectCount1` selector function in advance and pass it to `useStore` to optimize re-renders:

```
const selectCount1 = (state: StoreState) => state.count1;

const Counter1 = () => {
  const count1 = useStore(selectCount1);
  const inc1 = () => {
    useStore.setState(
      (prev) => ({ count1: prev.count1 + 1 })
    );
  };
  return (
    <div>
      count1: {count1} <button onClick={inc1}>+1</button>
    </div>
  );
};
```

Notice that the inline `inc1` function is defined. We invoke the `setState` function in `store`. This is a typical pattern and we can define the function in `store` for more reusability and readability.

The `store` creator function that is passed to the `create` function takes some arguments; the first argument is the `setState` function in `store`. Let's redefine our `store` with this capability:

```
type StoreState = {
  count1: number;
  count2: number;
  inc1: () => void;
  inc2: () => void;
};

const useStore = create<StoreState>((set) => ({
  count1: 0,
  count2: 0,
  inc1: () => set(
    (prev) => ({ count1: prev.count1 + 1 })
  ),
  inc2: () => set(
    (prev) => ({ count2: prev.count2 + 1 })
  ),
}));
```

Now, our `store` has two new properties called `inc1` and `inc2`, which are function properties. Note that it's a good convention to name the first argument `set`, which is short for `setState`.

Using the new `store`, we must define the `Counter2` component. You can compare it to the previous `Counter1` component and notice that it can be refactored in the same way:

```
const selectCount2 = (state: StoreState) => state.count2;
const selectInc2 = (state: StoreState) => state.inc2;

const Counter2 = () => {
  const count2 = useStore(selectCount2);
  const inc2 = useStore(selectInc2);
  return (
```

```
      <div>
        count2: {count2} <button onClick={inc2}>+1</button>
      </div>
    );
  };
```

In this example, we have a new selector function called `selectInc2`, and the `inc2` function is just the result of `useStore`. Likewise, we could add more functions to `store`, which allows some logic to reside outside the components. You can co-locate state updating logic close to the state values. This is the reason why Zustand's `setState` merges old state and new state. We also discussed this in the *Understanding module state and immutable state* section, where we learned how `Object.assign` is used.

What if we want to create a derived state? We can use a selector for a derived state. First, let's look at a naive example. The following is a new component that shows the `total` number of `count1` and `count2`:

```
  const Total = () => {
    const count1 = useStore(selectCount1);
    const count2 = useStore(selectCount2);
    return (
      <div>
        total: {count1 + count2}
      </div>
    );
  };
```

This is a valid pattern and it can stay as-is. There is an edge case where extra re-renders happen, which is when `count1` is increased and `count2` is decreased by the same amount. The `total` number won't change, but it will re-render. To avoid this, we can use a selector function for the derived state.

The following example shows a new `selectTotal` function being used to calculate the `total` number:

```
  const selectTotal =
    (state: StoreState) => state.count1 + state.count2;

  const Total = () => {
    const total = useStore(selectTotal);
    return (
```

```
    <div>
      total: {total}
    </div>
  );
};
```

This will only re-render when the `total` number is changed.

With that, we have calculated the `total` number in a selector. While this is a valid solution, let's look at another approach where we can create the total number in the store. If we could create the `total` number in `store`, it could remember the result and we can avoid unnecessary calculations when many components are using the value. This is not very common, but it's important if the calculation is very computation-heavy. A naive way to do this would be as follows:

```
const useStore = create((set) => ({
  count1: 0,
  count2: 0,
  total: 0,
  inc1: () => set((prev) => ({
    ...prev,
    count1: prev.count1 + 1,
    total: prev.count1 + 1 + prev.count2,
  })),
  inc2: () => set((prev) => ({
    ...prev,
    count2: prev.count2 + 1,
    total: prev.count2 + 1 + prev.count1,
  })),
}));
```

There is a more sophisticated way to do this, but the base idea is to calculate multiple properties at the same time and keep them in sync. Another library, Jotai, handles this well. Refer to *Chapter 8, Use Case Scenario 2 – Jotai*, to learn more.

The last missing piece for running the example app is the App component:

```
const App = () => (
  <>
    <Counter1 />
```

```
    <Counter2 />
    <Total />
  </>
);
```

When you run this app, you will see something like the following:

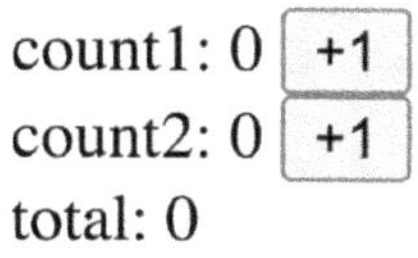

count1: 0 [+1]
count2: 0 [+1]
total: 0

Figure 7.1 – Screenshot of the running app

If you click the first button, you will see that both numbers on the screen – after the count1 label and the total number – increase. If you click the second button, you will see that both numbers on the screen – after the count2 label and the total number – increase.

In this section, we learned about reading and updating the state in a way that is often used in Zustand. Next, we will learn about how to handle structured data and how to use arrays.

Handling structured data

An example that deals with a set of numbers is fairly easy. In reality, we need to handle objects, arrays, and a combination of them. Let's learn how to use Zustand by covering another example. This is a well-known Todo app example. It's an app where you can do the following things:

- Create a new Todo item.
- See the list of Todo items.
- Toggle a Todo item's done status.
- Remove a Todo item.

First, we must define some types before creating a store. The following is the type definition for a Todo object. It has the id, title, and done properties:

```
type Todo = {
  id: number;
  title: string;
```

```
  done: boolean;
};
```

Now, the `StoreState` type can be defined with `Todo`. The value part of the store is `todos`, which is a list of Todo items. In addition to this, there are three functions – `addTodo`, `removeTodo`, and `toggleTodo` – that can be used to manipulate the `todos` property:

```
type StoreState = {
  todos: Todo[];
  addTodo: (title: string) => void;
  removeTodo: (id: number) => void;
  toggleTodo: (id: number) => void;
};
```

The `todos` property is an array of objects. Having an array of objects in a `store` state is a typical practice and will be the focus of this section.

Next, we must define `store`. It's also a hook that's called `useStore`. When it's created, `store` has an empty `todos` property and three functions called `addTodo`, `removeTodo`, and `toggleTodo`. `nextId` is defined outside the `create` function as a naive solution to provide a unique `id` for a new Todo item:

```
let nextId = 0;

const useStore = create<StoreState>((set) => ({
  todos: [],
  addTodo: (title) =>
    set((prev) => ({
      todos: [
        ...prev.todos,
        { id: ++nextId, title, done: false },
      ],
    })),
  removeTodo: (id) =>
    set((prev) => ({
      todos: prev.todos.filter((todo) => todo.id !== id),
    })),
  toggleTodo: (id) =>
```

```
    set((prev) => ({
      todos: prev.todos.map((todo) =>
        todo.id === id ? { ...todo, done: !todo.done } :
          todo
      ),
    })),
  }));
```

Notice that the `addTodo`, `removeTodo`, and `toggleTodo` functions are implemented in an immutable manner. They don't mutate existing objects and arrays; they create new ones instead.

Before we define a main `TodoList` component, let's look at a `TodoItem` component that is responsible for rendering one item:

```
const selectRemoveTodo =
  (state: StoreState) => state.removeTodo;
const selectToggleTodo =
  (state: StoreState) => state.toggleTodo;

const TodoItem = ({ todo }: { todo: Todo }) => {
  const removeTodo = useStore(selectRemoveTodo);
  const toggleTodo = useStore(selectToggleTodo);
  return (
    <div>
      <input
        type="checkbox"
        checked={todo.done}
        onChange={() => toggleTodo(todo.id)}
      />
      <span
        style={{
          textDecoration:
            todo.done ? "line-through" : "none",
        }}
      >
        {todo.title}
      </span>
```

```
      <button
        onClick={ () => removeTodo (todo.id) }
      >
        Delete
      </button>
    </div>
  );
};
```

As the `TodoItem` component takes a `todo` object in `props`, it's a fairly simple component in terms of states. The `TodoItem` component has two controls: a button that is handled by `removeTodo` and a checkbox that is handled by `toggleTodo`. These are the two functions from `store` for each control. The `selectRemoveTodo` and `selectToggleTodo` functions are passed to the `useStore` function to get the `removeTodo` and `toggleTodo` functions, respectively.

Let's create a memoized version of the `TodoItem` component named `MemoedTodoItem`:

```
const MemoedTodoItem = memo (TodoItem) ;
```

Now, we will discuss how this will help in our app. We are ready to define the main `TodoList` component. It uses `selectTodos`, a function that's used to select the `todos` property from `store`. Then, it maps over the `todos` array and renders `MemoedTodoItem` for each todo item.

It is important to use the memoized component here to avoid extra re-renders. Because we update the `store` state in an immutable manner, most of the `todo` objects in the `todos` array are not changed. If the `todo` object we pass to the `MemoedTodoItem` props is not changed, the component won't re-render. Whenever the `todos` array is changed, the `TodoList` component re-renders. However, its child components only re-render if the corresponding `todo` item is changed.

The following code shows the `selectTodos` function and the `TodoList` component:

```
const selectTodos = (state: StoreState) => state.todos;

const TodoList = () => {
  const todos = useStore (selectTodos) ;
  return (
    <div>
```

```
      {todos.map((todo) => (
        <MemoedTodoItem key={todo.id} todo={todo} />
      ))}
    </div>
  );
};
```

The `TodoList` component maps over the `todos` list and, for each `todo` item, renders the `MemoedTodoItem` component.

What remains is to add a new `todo` item. `NewTodo` is a component that can be used to render a text box and a button, as well as to call the `addTodo` function when the button is clicked. `selectAddTodo` is a function that can be used to select the `addTodo` function in `store`:

```
const selectAddTodo = (state: StoreState) => state.addTodo;

const NewTodo = () => {
  const addTodo = useStore(selectAddTodo);
  const [text, setText] = useState("");
  const onClick = () => {
    addTodo(text);
    setText(""); // [1]
  };
  return (
    <div>
      <input
        value={text}
        onChange={(e) => setText(e.target.value)}
      />
      <button onClick={onClick} disabled={!text}> // [2]
        Add
      </button>
    </div>
  );
};
```

There are two minor notes we should mention regarding improving the behavior in NewTodo:

- It clears the text box when the button is clicked [1].
- It disables the button when the text box is empty [2].

Finally, to finish up the Todo app, we must define the App component:

```
const App = () => (
  <>
    <TodoList />
    <NewTodo />
  </>
);
```

Running this app will show only a text box and a disabled **Add** button at first:

Figure 7.2 – First screenshot of the running app

If you enter some text and click the **Add** button, the item will appear:

Figure 7.3 – Second screenshot of the running app

Clicking a checkbox will toggle the done status of the item:

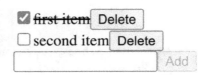

Figure 7.4 – Third screenshot of the running app

Clicking the **Delete** button on the screen will delete the item:

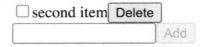

Figure 7.5 – Fourth screenshot of the running app

You can add as many items as you want. All these features are implemented with all the code we have discussed in this section. Re-renders are optimized, thanks to the immutable update of the `store` state and the `memo` function provided by React.

In this section, we learned how to handle arrays with a typical Todo app example. Next, we will discuss the pros and cons of this library and the approach in general.

Pros and cons of this approach and library

Let's discuss the pros and cons of using Zustand or any other libraries to implement this approach.

To recap, the following are the reading and writing states of Zustand:

- **Reading state**: This utilizes selector functions to optimize re-renders.
- **Writing state**: This is based on the immutable state model.

The key point is that React is based on object immutability for optimization. One example is useState. React optimizes re-renders with object referential equality based on immutability. The following example illustrates this behavior:

```
const countObj = { value: 0 };

const Component = () => {
  const [count, setCount] = useState(countObj);
  const handleClick = () => {
    setCount(countObj);
  };
  useEffect(() => {
    console.log("component updated");
  });
  return (
    <>
      {count.value}
      <button onClick={handleClick}>Update</button>
    </>
  );
};
```

Here, even if you click the Update button, it won't show the "component updated" message. This is because React assumes that the countObj value will not change if the object reference is the same. This means that changing the handleClick function doesn't make any changes:

```
const handleClick = () => {
  countObj.value += 1;
  setCount(countObj);
};
```

If you call handleClick, the countObj value will change, but the countObj object won't. Hence, React assumes it's unchanged. This is what we mean by React being based on immutability for optimization. This same behavior can be observed with functions such as memo and useMemo.

The Zustand state model is perfectly in line with this object immutability assumption (or convention). Zustand's render optimization with selector functions is also based on immutability – that is, if a selector function returns the same object referentially (or value), it assumes that the object is not changed and avoids re-rendering.

Zustand having the same model as React gives us a huge benefit in terms of library simplicity and its small bundle size.

On the other hand, a limitation of Zustand is its manual render optimization with selectors. It requires that we understand object referential equality and the code for selectors tends to require more boilerplate code.

In summary, Zustand – or any other libraries with this approach – is a simple addition to the React principle. It's a good recommendation if you need a library with a small bundle size, if you are familiar with referential equality and memoization, or you prefer manual render optimization.

Summary

In this chapter, we learned about the Zustand library. It's a tiny library that uses module state in React. We looked at a counting example and a Todo example to grasp how to use the library. We typically use this library to understand object referential equality. You can choose this library or similar approaches based on your requirements and what you have learned in this chapter.

We didn't discuss some aspects of Zustand in this chapter, including middleware, which allows you to give some features to the `store` creator, and non-module state usage, which creates a `store` in the React life cycle. These can be other considerations when you're choosing a library. You should always refer to the library documentation for more – and the latest – information.

In the next chapter, we will learn about another library, Jotai.

8
Use Case Scenario 2 – Jotai

Jotai (`https://github.com/pmndrs/jotai`) is a small library for the global state. It's modeled after `useState`/`useReducer` and with what are called atoms, which are usually small pieces of state. Unlike Zustand, it is a component state, and like Zustand, it is an immutable update model. The implementation is based on the Context and Subscription patterns we learned about in *Chapter 5, Sharing Component State with Context and Subscription*.

In this chapter, we will learn about the basic usage of the Jotai library and how it deals with optimizing re-renders. With atoms, the library can track dependencies and trigger re-renders based on the dependencies. Because Jotai internally uses Context and atoms themselves do not hold values, atom definitions are reusable, unlike the module state. We will also discuss a novel pattern with atoms, called **Atoms-in-Atom**, which is a technique to optimize re-renders with an array structure.

In this chapter, we will cover the following topics:

- Understanding Jotai
- Exploring render optimization
- Understanding how Jotai works to store atom values

- Adding an array structure
- Using the different features of Jotai

Technical requirements

You are expected to have moderate knowledge of React, including React hooks. Refer to the official site, `https://reactjs.org`, to learn more.

In some code, we use TypeScript (`https://www.typescriptlang.org`), and you should have basic knowledge of it.

The code in this chapter is available on GitHub at `https://github.com/PacktPublishing/Micro-State-Management-with-React-Hooks/tree/main/chapter_08`.

To run the code snippets in this chapter, you need a React environment—for example, Create React App (`https://create-react-app.dev`) or CodeSandbox (`https://codesandbox.io`).

Understanding Jotai

To understand the Jotai **application programming interface** (**API**), let's remind ourselves of a simple counter example and the solution with Context.

Here is an example with two separate counters:

```
const Counter1 = () => {
  const [count, setCount] = useState(0); // [1]
  const inc = () => setCount((c) => c + 1);
  return <>{count} <button onClick={inc}>+1</button></>;
};

const Counter2 = () => {
  const [count, setCount] = useState(0);
  const inc = () => setCount((c) => c + 1);
  return <>{count} <button onClick={inc}>+1</button></>;
};

const App = () => (
  <>
```

```
    <div><Counter1 /></div>
    <div><Counter2 /></div>
  </>
);
```

Because these `Counter1` and `Counter2` components have their own local states, the numbers shown in these components are isolated.

If we want those two components to share a single count state, we can lift the state up and use Context to pass it down, as we discussed in the *Effectively using local states* section of *Chapter 2, Using Local and Global States*. Let's see an example that is solved with Context.

First, we create a `Context` variable to hold the count state, as follows:

```
const CountContext = createContext();

const CountProvider = ({ children }) => (
  <CountContext.Provider value={useState(0)}>
    {children}
  </CountContext.Provider>
);
```

Notice the `Context` value is the same state, `useState(0)`, as we used in the previous example (marked **[1]**).

Then, the following are the modified components, where we replace `useState(0)` with `useContext(CountContext)`:

```
const Counter1 = () => {
  const [count, setCount] = useContext(CountContext);
  const inc = () => setCount((c) => c + 1);
  return <>{count} <button onClick={inc}>+1</button></>;
};

const Counter2 = () => {
  const [count, setCount] = useContext(CountContext);
  const inc = () => setCount((c) => c + 1);
  return <>{count} <button onClick={inc}>+1</button></>;
};
```

Finally, we wrap those components with `CountProvider`, like this:

```
const App = () => (
  <CountProvider>
    <div><Counter1 /></div>
    <div><Counter2 /></div>
  </CountProvider>
);
```

This makes it possible to have a shared count state, and you will see that two `count` numbers in `Counter1` and `Counter2` components are incremented at once.

Now, let's see how Jotai is helpful compared to Context. There are two benefits when using Jotai, as follows:

- Syntax simplicity
- Dynamic atom creation

Let's start with the first benefit—how Jotai can help to simplify the syntax.

Syntax simplicity

To understand syntax simplicity, let's look at the same counter example with Jotai. First, we need to import some functions from the Jotai library, as follows:

```
import { atom, useAtom } from "jotai";
```

The `atom` function and the `useAtom` hook are basic functions provided by Jotai.

An atom represents a piece of a state. An atom is usually a small piece of state, and it is a minimum unit of triggering re-renders. The `atom` function creates a definition of an atom. The `atom` function takes one argument to specify an initial value, just as `useState` does. The following code is used to define a new atom:

```
const countAtom = atom(0);
```

Notice the similarity with `useState(0)`.

Now, we use the atom in counter components. Instead of useState(0), we use
useAtom(countAtom), as follows:

```
const Counter1 = () => {
  const [count, setCount] = useAtom(countAtom);
  const inc = () => setCount((c) => c + 1);
  return <>{count} <button onClick={inc}>+1</button></>;
};


const Counter2 = () => {
  const [count, setCount] = useAtom(countAtom);
  const inc = () => setCount((c) => c + 1);
  return <>{count} <button onClick={inc}>+1</button></>;
};
```

Because useAtom(countAtom) returns the same tuple, [count, setCount],
as useState(0) does, the rest of the code doesn't need to be changed.

Finally, our App component is the same as in the first example of this chapter, which is
without Context, as illustrated in the following code snippet:

```
const App = () => (
  <>
    <div><Counter1 /></div>
    <div><Counter2 /></div>
  </>
);
```

Unlike the second example of this chapter, which is with Context, we don't need
a provider. This is possible due to the "default store" in Context, as we learned in
the *Implementing the Context and Subscription pattern* section of *Chapter 5, Sharing
Component State with Context and Subscription*. We can optionally use a provider when
we need to provide different values for different subtrees.

To have a better understanding of the syntax simplicity in Jotai, let's suppose you want to add another global state—say, `text`; you would end up adding the following code:

```
const TextContext = createContext();

const TextProvider = ({ children }) => (
  <TextContext.Provider value={useState("")}>
    {children}
  </TextContext.Provider>
);

const App = () => (
  <TextProvider>

    ...

  </TextProvider>
);

// When you use it in a component
  const [text, setText] = useContext(TextContext);
```

This is not too bad. What we added is a Context definition and a provider definition, and we wrapped `App` with the `Provider` component. You can also avoid provider nesting, as we learned in the *Best practices for using Context* section of *Chapter 3, Sharing the Component State with Context.*

However, the same example could be done with Jotai atoms, as follows:

```
const textAtom = atom("");

// When you use it in a component
  const [text, setText] = useAtom(textAtom);
```

This is far simpler. Essentially, we added just a one-line atom definition. Even if we had more atoms, we would just need a line for each atom definition in Jotai. On the other hand, using Context would require creating a Context for each piece of state. It's possible to do it with Context, but not trivial. Jotai's syntax is much more simplified. This is the first benefit of Jotai.

While the syntax simplicity is great, it doesn't give any new capability. Let's briefly discuss the second benefit.

Dynamic atom creation

The second benefit of Jotai is a new capability—that is, dynamic atom creation. Atoms can be created and destroyed in the React component lifecycle. This is not possible with the multiple-Context approach, because adding a new state means adding a new `Provider` component. If you add a new component, all its child components will be remounted, throwing away their states. We will cover a use case of dynamic atom creation in the *Adding an array structure* section.

The implementation of Jotai is based on what we learned in *Chapter 5, Sharing Component State with Context and Subscription*. Jotai's store is basically a `WeakMap` object (`https://developer.mozilla.org/en-US/docs/Web/JavaScript/Reference/Global_Objects/WeakMap`) of atom config objects and atom values. An **atom config object** is a definition created with the `atom` function. An **atom value** is a value that the `useAtom` hook returns. Subscription in Jotai is atom-based, which means the `useAtom` hook subscribes to a certain atom in `store`. Atom-based Subscription gives the ability to avoid extra re-renders. We will discuss this further in the next section.

In this section, we discussed the basic mental model and the API of the Jotai library. Next up, we will dive into how the atom model solves render optimization.

Exploring render optimization

Let's recap on selector-based render optimization. We will start by using an example from *Chapter 4, Sharing Module State with Subscription*, where we created `createStore` and `useStoreSelector`.

Let's define a new `store` person with `createStore`. We define three properties: `firstName`, `lastName`, and `age`, as follows:

```
const personStore = createStore({
  firstName: "React",
  lastName: "Hooks",
  age: 3,
});
```

Suppose we would like to create a component that shows `firstName` and `lastName`. One straightforward way is to select those properties. Here is an example with `useStoreSelector`:

```
const selectFirstName = (state) => state.firstName;
const selectLastName = (state) => state.lastName;

const PersonComponent = () => {
  const firstName =
    useStoreSelector(store, selectFirstName);
  const lastName = useStoreSelector(store, selectLastName);
  return <>{firstName} {lastName}</>;
};
```

As we have selected only two properties from the `store`, when the non-selected property, age, is changed, `PersonComponent` will not re-render.

This `store` and selector approach is what we call **top-down**. We create a `store` that holds everything and select pieces of state from the `store` as necessary.

Now, what would Jotai atoms look like for the same example? First, we define atoms, as follows:

```
const firstNameAtom = atom("React");
const lastNameAtom = atom("Hooks");
const ageAtom = atom(3);
```

Atoms are units of triggering re-renders. You can make atoms as small as you want to control re-renders, like primitive values. But atoms can be objects too.

`PersonComponent` can be implemented with the `useAtom` hook, as follows:

```
const PersonComponent = () => {
  const [firstName] = useAtom(firstNameAtom);
  const [lastName] = useAtom(lastNameAtom);
  return <>{firstName} {lastName}</>;
};
```

Because this has no relationship with `ageAtom`, `PersonComponent` won't re-render when the value of `ageAtom` is changed.

Atoms can be as small as possible, but that means we would probably have too many atoms to organize. Jotai has a notion of derived atoms, where you can create another atom from existing atoms. Let's create a `personAtom` variable that holds the first name, last name, and age. We can use the `atom` function, which takes a `read` function to generate a derived value. The code is illustrated in the following snippet:

```
const personAtom = atom((get) => ({
  firstName: get(firstNameAtom),
  lastName: get(lastNameAtom),
  age: get(ageAtom),
}));
```

The `read` function takes an argument called `get`, with which you can refer to other atoms and get their values. The value of `personAtom` is an object with three properties— `firstName`, `lastName`, and `age`. This value is updated whenever one of the properties is changed, which means when `firstNameAtom`, `lastNameAtom`, or `ageAtom` is updated. This is called dependency tracking and is automatically done by the Jotai library.

> **Important Note**
> Dependency tracking is dynamic and works for conditional evaluations. For example, suppose a `read` function is `(get) => get(a) ? get(b) : get(c)`. In this case, if the value of a is truthy, the dependency is a and b, whereas if the value of a is falsy, the dependency is a and c.

Using `personAtom`, we could re-implement `PersonComponent`, as follows:

```
const PersonComponent = () => {
  const person = useAtom(personAtom);
  return <>{person.firstName} {person.lastName}</>;
};
```

However, this is not what we expect. It will re-render when `ageAtom` changes its value, hence causing extra re-renders.

To avoid extra re-renders, we should create a derived atom including only values we use. Here is another atom, named `fullNameAtom` this time:

```
const fullNameAtom = atom((get) => ({
  firstName: get(firstNameAtom),
  lastName: get(lastNameAtom),
}));
```

Using `fullNameAtom`, we can implement `PersonComponent` once again, like this:

```
const PersonComponent = () => {
  const person = useAtom(fullNameAtom);
  return <>{person.firstName} {person.lastName}</>;
};
```

Thanks to `fullNameAtom`, this doesn't re-render even when the `ageAtom` value is changed.

We call this a **bottom-up** approach. We create small atoms and combine them to create bigger atoms. We can optimize re-renders by adding only atoms that will be used in components. The optimization is not automatic, but more straightforward with the atom model.

How could we do the last example with a store and selector approach? Here is an example with an `identity` selector:

```
const identity = (x) => x;

const PersonComponent = () => {
  const person = useStoreSelector(store, identity);
  return <>{person.firstName} {person.lastName}</>;
};
```

As you might guess, this causes extra re-renders. When the `age` property in the `store` is changed, the component re-renders.

A possible fix would be to select only `firstName` and `lastName`. The following example illustrates this:

```
const selectFullName = (state) => ({
  firstName: state.firstName,
  lastName: state.lastName,
});

const PersonComponent = () => {
  const person = useStoreSelector(store, selectFullName);
  return <>{person.firstName} {person.lastName}</>;
};
```

Unfortunately, this doesn't work. When `age` is changed, the `selectFullName` function is re-evaluated, and it returns a new object with the same property values. `useStoreSelector` assumes the new object may contain new values and trigger re-renders, which causes extra re-renders. This is a well-known issue with the selector approach, and typical solutions are to use either a custom equality function or a memoization technique.

The benefit of the atom model is that the composition of atoms can easily relate to what will be shown in a component. Thus, it's straightforward to control re-renders. Render optimization with atoms doesn't require the custom equality function or the memoization technique.

Let's look at a counter example to learn more about the derived atoms. First, we define two `count` atoms, as follows:

```
const count1Atom = atom(0);
const count2Atom = atom(0);
```

We define a component to use those `count` atoms. Instead of defining two counter components, we define a single `Counter` component that works for both atoms. To this end, the component receives `countAtom` in its `props`, as illustrated in the following code snippet:

```
const Counter = ({ countAtom }) => {
  const [count, setCount] = useAtom(countAtom);
  const inc = () => setCount((c) => c + 1);
  return <>{count} <button onClick={inc}>+1</button></>;
};
```

This is reusable for any `countAtom` configs. Even if we define a new `count3Atom` config, we don't need to define a new component.

Next, we define a derived atom that calculates the total number of two counts. We use `atom` with a `read` function as the first argument, as follows:

```
const totalAtom = atom(
  (get) => get(count1Atom) + get(count2Atom)
);
```

With the `read` function, `atom` will create a derived atom. The value of the derived atom is the result of the `read` function. The derived atom will re-evaluate its `read` function and update its value only when dependencies are changed. In this case, either `count1Atom` or `count2Atom` is changed.

The `Total` component is a component to use `totalAtom` and show the `total` number, as illustrated in the following code snippet:

```
const Total = () => {
  const [total] = useAtom(totalAtom);
  return <>{total}</>;
};
```

`totalAtom` is a derived atom and it's read-only because its value is the result of the `read` function. Hence, there's no notion of setting a value of `totalAtom`.

Finally, we define an `App` component. It passes `count1Atom` and `count2Atom` to `Counter` components, as follows:

```
const App = () => (
  <>
    (<Counter countAtom={count1Atom} />)
    +
    (<Counter countAtom={count2Atom} />)
    =
    <Total />
  </>
);
```

Atoms can be passed as `props`, such as the `Counter` atom in this example, or they can be passed by any other means—constants at the module level, `props`, contexts, or even as values in other atoms. We will learn about the use case of putting atoms in another atom in the *Adding an array structure* section.

When you run the app, you will see an equation of the first count, the second count, and the total number. By clicking the buttons shown right after the counts, you will see the count incremented as well as the total number, as illustrated in the following screenshot:

$$(2 \boxed{+1}) + (3 \boxed{+1}) = 5$$

Figure 8.1 – Screenshot of the counter app

In this section, we learned about the atom model and render optimization in the Jotai library. Next up, we'll look into how Jotai stores atom values.

Understanding how Jotai works to store atom values

So far, we haven't discussed how Jotai uses Context. In this section, we'll show how Jotai stores atom values and how atoms are reusable.

First, let's revisit a simple atom definition, `countAtom`. `atom` takes an initial value of 0 and returns an atom config, as follows:

```
const countAtom = atom(0);
```

Implementation-wise, `countAtom` is an object holding some properties representing the atom behavior. In this case, `countAtom` is a primitive atom, which is an atom with a value that can be updated with a value or an updating function. A primitive atom is designed to behave like `useState`.

What is important is that atom configs such as `countAtom` don't hold their values. We have a `store` that holds atom values. A `store` has a `WeakMap` object whose key is an atom config object and whose value is an atom value.

When we use `useAtom`, by default, it uses a default `store` defined at the module level. However, Jotai provides a component named `Provider`, which lets you create a `store` at the component level. We can import `Provider` from the Jotai library along with `atom` and `useAtom`, as follows:

```
import { atom, useAtom, Provider } from "jotai";
```

Let's suppose we have the `Counter` component defined, as follows:

```
const Counter = ({ countAtom }) => {
  const [count, setCount] = useAtom(countAtom);
  const inc = () => setCount((c) => c + 1);
  return <>{count} <button onClick={inc}>+1</button></>;
};
```

This is the same component we defined in the *Understanding Jotai* section and the *Exploring render optimization* section.

We then define an App component using Provider. We use two Provider components and put in two Counter components for each Provider component, as follows:

```
const App = () => (
  <>
    <Provider>
      <h1>First Provider</h1>
      <div><Counter /></div>
      <div><Counter /></div>
    </Provider>
    <Provider>
      <h1>Second Provider</h1>
      <div><Counter /></div>
      <div><Counter /></div>
    </Provider>
  </>
);
```

The two Provider components in App isolate stores. Hence, countAtom used in Counter components is isolated. The two Counter components under the first Provider component share the countAtom value, but the other two Counter components under the second Provider component have different values of countAtom from the value in the first Provider component, as shown here:

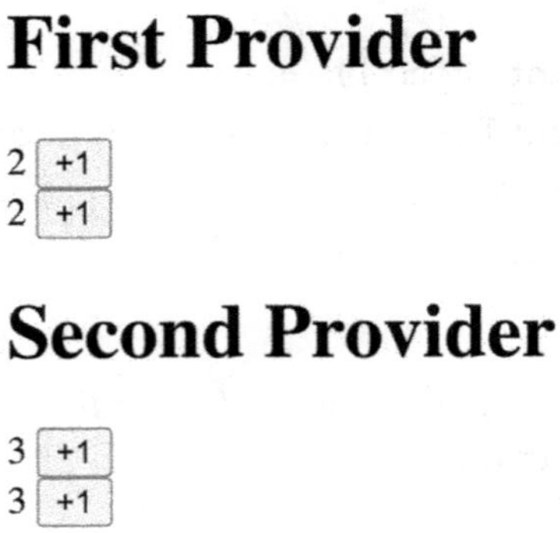

First Provider

2 +1
2 +1

Second Provider

3 +1
3 +1

Figure 8.2 – Screenshot of the two-provider app

Again, what is important is that countAtom itself doesn't hold a value. Thus, countAtom is reusable for multiple Provider components. This is a notable difference from module states.

We could define a derived atom. Here is a derived atom to define the doubled number of `countAtom`:

```
const doubledCountAtom = atom(
  (get) => get(countAtom) * 2
);
```

As `countAtom` doesn't hold a value, `doubledCountAtom` doesn't either. If `doubledCountAtom` is used in the first `Provider` component, it represents the doubled value of the `countAtom` value in the `Provider` component. The same applies to the second `Provider` component, and the values in the first `Provider` component can be different from the values in the second `Provider` component.

Because atom configs are just definitions that don't hold values, the atom configs are reusable. The example shows it's reusable for two `Provider` components, but essentially, it's reusable for more `Provider` components. Furthermore, a `Provider` component can be used dynamically in the React component life cycle. Implementation-wise, Jotai is totally based on Context, and Jotai can do everything that Context can do. In this section, we learned that atom configs don't hold values and thus are reusable. Next up, we will learn how to deal with arrays with Jotai.

Adding an array structure

An array structure is tricky to handle in React. When a component renders an array structure, we need to pass stable `key` properties to the array items. This is especially necessary when we remove or reorder the array items.

In this section, we'll learn how to handle array structures in Jotai. We'll start with a traditional approach, and then a new pattern that we call **Atoms-in-Atom**.

Let's use the same to-do app example that we used in the *Handling structured data* section of *Chapter 7, Use Case Scenario 1 – Zustand*.

First, we define a `Todo` type. It has the `id` string, `title` string, and `done` Boolean properties, as illustrated in the following code snippet:

```
type Todo = {
  id: string;
  title: string;
  done: boolean;
};
```

Next, we define `todosAtom`, which represents an array of defined `Todo` items, as follows:

```
const todosAtom = atom<Todo[]>([]);
```

We annotate the `atom()` function with the `Todo[]` type.

We then define a `TodoItem` component. This is a pure component that receives `todo`, `removeTodo`, and `toggleTodo` as props. The code is illustrated in the following snippet:

```
const TodoItem = ({
  todo,
  removeTodo,
  toggleTodo,
}: {
  todo: Todo;
  removeTodo: (id: string) => void;
  toggleTodo: (id: string) => void;
}) => {
  return (
    <div>
      <input
        type="checkbox"
        checked={todo.done}
        onChange={() => toggleTodo(todo.id)}
      />
      <span
        style={{
          textDecoration:
            todo.done ? "line-through" : "none",
        }}
      >
        {todo.title}
      </span>
      <button
        onClick={() => removeTodo(todo.id)}
      >Delete</button>
    </div>
```

```
    );
  };
```

The onChange callback in <input> invokes toggleTodo, and the onClick callback in <button> invokes removeTodo. Both are based on the id string.

We wrap TodoItem with memo to create a memoized version, as follows:

```
const MemoedTodoItem = memo(TodoItem);
```

This allows us to avoid re-renders unless todo, removeTodo, or toggleTodo are changed.

Now, we are ready to create a TodoList component. It uses todosAtom, defines removeTodo and toggleTodo with useCallback, and maps over the todo array, as follows:

```
const TodoList = () => {
  const [todos, setTodos] = useAtom(todosAtom);
  const removeTodo = useCallback((id: string) => setTodos(
    (prev) => prev.filter((item) => item.id !== id)
  ), [setTodos]);
  const toggleTodo = useCallback((id: string) => setTodos(
    (prev) => prev.map((item) =>
      item.id === id ? { ...item, done: !item.done } : item
    )
  ), [setTodos]);
  return (
    <div>
      {todos.map((todo) => (
        <MemoedTodoItem
          key={todo.id}
          todo={todo}
          removeTodo={removeTodo}
          toggleTodo={toggleTodo}
        />
      ))}
    </div>
  );
};
```

The `TodoList` component renders the `MemoedTodoItem` component for each `todos` array item. The `key` prop is specified as `todo.id`.

The next component is `NewTodo`. It uses `todosAtom` and adds a new item on button click. The `id` value of the new atom should be uniquely generated, and in the following example, it uses `nanoid` (`https://www.npmjs.com/package/nanoid`):

```
const NewTodo = () => {
  const [, setTodos] = useAtom(todosAtom);
  const [text, setText] = useState("");
  const onClick = () => {
    setTodos((prev) => [
      ...prev,
      { id: nanoid(), title: text, done: false },
    ]);
    setText("");
  };
  return (
    <div>
      <input
        value={text}
        onChange={(e) => setText(e.target.value)}
      />
      <button onClick={onClick} disabled={!text}>
        Add
      </button>
    </div>
  );
};
```

For simplicity, we used `useAtom` for `todosAtom`. However, this actually makes the `NewTodo` component re-render when the value of `todosAtom` is changed. We could easily avoid this with an additional utility hook called `useUpdateAtom`.

Finally, we create an `App` component to render `TodoList` and `NewTodo`, as follows:

```
const App = () => (
  <>
    <TodoList />
```

```
    <NewTodo />
  </>
);
```

This works perfectly. You can add, remove, and toggle to-do items without any issues, as shown here:

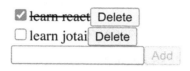

Figure 8.3 – Screenshot of the Todo app

There are two concerns, though, from the developer's perspective, as follows:

- The first concern is we need to modify the entire `todos` array to mutate a single item. In the `toggleTodo` function, it needs to iterate over all the items and mutate just one item. In the atomic model, it would be nice if we could simply mutate one item. This is also related to performance. When `todos` array items are mutated, the `todos` array itself is changed. Thus, `TodoList` re-renders. Thanks to `MemoedTodoItem`, the `MemoedTodoItem` components don't re-render unless the specific item is changed. Ideally, we want to trigger those specific `MemoedTodoItem` components to re-render.

- The second concern is the `id` value of an item. The `id` value is primarily for `key` in `map`, and it would be nice if we could avoid using `id`.

With Jotai, we propose a new pattern, **Atoms-in-Atom**, with which we put atom configs in another atom value. This pattern addresses the two concerns and is more consistent with Jotai's mental model.

Let's see how we can re-create the same Todo app we created previously in this section with the new pattern.

We start by defining the `Todo` type, as follows:

```
type Todo = {
  title: string;
  done: boolean;
};
```

This time, the `Todo` type doesn't have an `id` value.

We then create a `TodoAtom` type with `PrimitiveAtom`, which is a generic type exported by the Jotai library. The code is illustrated in the following snippet:

```
type TodoAtom = PrimitiveAtom<Todo>;
```

We use this `TodoAtom` type to create a `todoAtomsAtom` config, as follows:

```
const todoAtomsAtom = atom<TodoAtom[]>([]);
```

The name is explicit, to tell that it's an `atom` that represents an array of `TodoAtom`. This structure is why the pattern is named **Atoms-in-Atom**.

Here is the `TodoItem` component. It receives `todoAtom` and `remove` properties. The component uses the `todoAtom` atom with `useAtom`:

```
const TodoItem = ({
  todoAtom,
  remove,
}: {
  todoAtom: TodoAtom;
  remove: (todoAtom: TodoAtom) => void;
}) => {
  const [todo, setTodo] = useAtom(todoAtom);
  return (
    <div>
      <input
        type="checkbox"
        checked={todo.done}
        onChange={() => setTodo(
          (prev) => ({ ...prev, done: !prev.done })
        )}
      />
      <span
        style={{
          textDecoration:
            todo.done ? "line-through" : "none",
        }}
      >
```

```
        {todo.title}
      </span>
      <button onClick={() => remove(todoAtom)}>
        Delete
      </button>
    </div>
  );
};
```

```
const MemoedTodoItem = memo(TodoItem);
```

Thanks to the useAtom config in the TodoItem component, the onChange callback is very simple and only cares about the item. It doesn't depend on the fact that it's an item of the array.

The TodoList component should be carefully looked at. It uses todoAtomsAtom, which returns todoAtoms as its value. The todoAtoms variable holds an array of todoAtom. The remove function is interesting as it takes todoAtom as the atom config and filters the todoAtom array in todoAtomsAtom. The full code of TodoList is shown here:

```
const TodoList = () => {
  const [todoAtoms, setTodoAtoms] =
    useAtom(todoAtomsAtom);
  const remove = useCallback(
    (todoAtom: TodoAtom) => setTodoAtoms(
      (prev) => prev.filter((item) => item !== todoAtom)
    ),
    [setTodoAtoms]
  );
  return (
    <div>
      {todoAtoms.map((todoAtom) => (
        <MemoedTodoItem
          key={`${todoAtom}`}
          todoAtom={todoAtom}
          remove={remove}
        />
```

```
    ))}
  </div>
 );
};
```

`TodoList` maps over the `todoAtoms` variable and renders `MemoedTodoItem` for each `todoAtom` config. For `key` in map, we specify the stringified `todoAtom` config. An atom config returns a **unique identifier (UID)** when evaluated as a string, thus we don't need to manage string IDs by ourselves. The behavior of the `TodoList` component is slightly different from the previous version. Because it deals with **Atoms-in-Atom**, `todoAtomsAtom` won't be changed if one of the items is toggled with `toggleTodo`. Thus, it can reduce some extra re-renders by nature.

The `NewTodo` component is almost the same as the previous example. One exception is that when creating a new item, it will create a new atom config and push it into `todoAtomsAtom`. The following snippet shows the `NewTodo` component code:

```
const NewTodo = () => {
  const [, setTodoAtoms] = useAtom(todoAtomsAtom);
  const [text, setText] = useState("");
  const onClick = () => {
    setTodoAtoms((prev) => [
      ...prev,
      atom<Todo>({ title: text, done: false }),
    ]);
    setText("");
  };
  return (
    <div>
      <input
        value={text}
        onChange={(e) => setText(e.target.value)}
      />
      <button onClick={onClick} disabled={!text}>
        Add
      </button>
    </div>
  );
};
```

The reset of the code and the behavior of the `NewTodo` component are basically equivalent to the previous example.

Finally, we have the same `App` component to run the app, as illustrated here:

```
const App = () => (
  <>
    <TodoList />
    <NewTodo />
  </>
);
```

If you run the app, you will see no differences from the previous example. As described, the differences are for developers.

Let's summarize the difference with the **Atoms-in-Atom** pattern, as follows:

- An array atom is used to hold an array of item atoms.
- To add a new item in the array, we create a new atom and add it.
- Atom configs can be evaluated as strings, and they return UIDs.
- A component that renders an item uses an item atom in each component. It eases mutating the item value and avoids extra re-renders naturally.

In this section, we learned how to handle the array structure. We saw two patterns—a naive one and an **Atoms-in-Atom** one—and their differences. Next up, we will learn about some other features that the Jotai library provides.

Using the different features of Jotai

So far, we've learned some basics of the Jotai library. There are some more basic features that we will cover in this section. These features are necessary if you need to deal with complex scenarios. We'll also briefly introduce some advanced features whose use cases are out of the scope of this book.

In this section, we'll discuss the following topics:

- Defining the `write` function of atoms
- Using action atoms
- Understanding the `onMount` option of atoms
- Introducing the `jotai/utils` bundle

- Understanding library usage
- Introduction to more advanced features

Let's take a look at each one now.

Defining the write function of atoms

We have seen how to create a derived atom. For example, `doubledCountAtom` with `countAtom` is defined in the *Understanding how Jotai works to store atom values* section, as follows:

```
const countAtom = atom(0);

const doubledCountAtom = atom(
  (get) => get(countAtom) * 2
);
```

`countAtom` is called a primitive atom because it's not derived from another atom. A primitive atom is a writable atom where you can change the value.

`doubledCountAtom` is a read-only derived atom because its value is fully dependent on `countAtom`. The value of `doubledCountAtom` can only be changed by changing the value of `countAtom`, which is a writable atom.

To create a writable derived atom, the `atom` function accepts an optional second argument for the `write` function, in addition to the first argument `read` function.

For example, let's redefine `doubledCountAtom` to be writable. We pass a `write` function that will change the value of `countAtom`, as follows:

```
const doubledCountAtom = atom(
  (get) => get(countAtom) * 2,
  (get, set, arg) => set(countAtom, arg / 2)
);
```

The `write` function takes three arguments, as follows:

- `get` is a function to return the value of an atom.
- `set` is a function to set the value of an atom.
- `arg` is an arbitrary value to receive when updating the atom (in this case, `doubledCountAtom`).

With the `write` function, the created atom is writable as if it is a primitive atom. Actually, it is not exactly the same as `countAtom` because `countAtom` accepts an updating function such as `setCount((c) => c + 1)`.

We can technically create a new atom that behaves identically to `countAtom`. What would be the use case? For example, you can add logging, as follows:

```
const anotherCountAtom = atom(
  (get) => get(countAtom),
  (get, set, arg) => {
    const nextCount = typeof arg === 'function' ?
      arg(get(countAtom)) : arg
    set(countAtom, nextCount)
    console.log('set count', nextCount)
  )
);
```

`anotherCountAtom` works like `countAtom`, and it shows a logging message when it sets a value.

Writable derived atoms are a powerful feature that can help in some complex scenarios. In the next subsection, we'll see another pattern using `write` functions.

Using action atoms

To organize state mutation code, we often create a function or a set of functions. We can use atoms for that purpose and call them action atoms.

To create action atoms, we only use the `write` function of the `atom` function's second argument. The first argument can be anything, but we often use `null` as a convention.

Let's look at an example. We have `countAtom` as usual and `incrementCountAtom`, which is an action atom, as follows:

```
const countAtom = count(0);

const incrementCountAtom(
  null,
  (get, set, arg) => set(countAtom, (c) => c + 1)
);
```

In this case, the `write` function of `incrementCountAtom` only uses `set`, out of three arguments.

We can use this atom like normal atoms, and just ignore its value. For example, here is a component to show a button to increment the count:

```
const IncrementButton = () => {
  const [, incrementCount] = useAtom(incrementCountAtom);
  return <button onClick={incrementCount}>Click</button>;
};
```

This is a simple case without an argument. You could accept an argument and you could create as many action atoms as you want.

Next, we will see a less commonly used but important feature.

Understanding the onMount option of atoms

In some use cases, we want to run certain logic once an atom starts to be used. A good example is to subscribe to an external data source. This can be done with the `useEffect` hook, but to define logic at the atom level, Jotai atoms have the `onMount` option.

To understand how it is used, let's create an atom that shows a login message on mount and unmount, as follows:

```
const countAtom = atom(0);
countAtom.onMount = (setCount) => {
  console.log("count atom starts to be used");
  const onUnmount = () => {
    console.log("count atom ends to be used");
  };
  return onUnmount;
};
```

The body of the `onMount` function is showing a logging message about the start of use. It also returns an `onUnmount` function, which shows a logging message about the end of use. The `onMount` function takes an argument, which is a function to update `countAtom`.

This is a contrived example, but there are many real use cases to connect external data sources.

Next, we'll talk about utility functions.

Introducing the jotai/utils bundle

The Jotai library provides two basic functions, `atom` and `useAtom`, and an additional `Provider` component in the main bundle. While the small API is good to understand the basic features, we want some utility functions to help development.

Jotai provides a separate bundle named `jotai/utils` that contains a variety of utility functions. For example, `atomWithStorage` is a function to create atoms with a specific feature—that is, to synchronize with persistent storage. For more information and other utility functions, refer to the project site at `https://github.com/pmndrs/jotai`.

Next, we will discuss how the Jotai library can be used in other libraries.

Understanding library usage

Suppose two libraries use the Jotai library internally. If we develop an app that uses the two libraries, there's an issue of double providers. Because Jotai atoms are distinguished by reference, it is possible that the atoms in the first library accidentally connect to the provider in the second library. As a result, it may not work as expected by the library authors. The Jotai library provides a notion of "scope", which is the way to connect to a specific provider. To make it work as expected, we should pass the same scope variable to the `Provider` component and the `useAtom` hook.

Implementation-wise, this is how Context works. The scope feature is just used to put back the Context feature. It's still under exploration how this feature can be used for other purposes. We, as a community, will work on more use cases with this feature.

Finally, we'll see some advanced features in the Jotai library.

Introduction to more advanced features

There are more advanced features that we didn't cover in this book.

Most notably, Jotai supports the React Suspense feature. When a derived atom's `read` function returns a promise, the `useAtom` hook will suspend, and React will show a fallback. This feature is experimental and subject to change, but it's a very important feature to explore.

Another note is about library integrations. Jotai is a library to solve a single problem with the atomic model, which is to avoid extra re-renders. By integrating with other libraries, the use case expands. The atomic model is flexible to integrate with other libraries, and especially, the `onMount` option is necessary for external data sources.

To learn more about these advanced features, refer to the project site:

`https://github.com/pmndrs/jotai`

In this section, we discussed some additional features that the Jotai library provides. Jotai is a primitive library to provide building blocks, yet is flexible enough to cover real use cases.

Summary

In this chapter, we learned about a library called Jotai. It's based on the atomic model and Context. We've seen simple examples to learn its basics, yet they show the flexibility of the atomic model. The combination of Context and Subscription is the only way to have a React-oriented global state. If your requirement is Context without extra re-renders, this approach should be your choice.

In the next chapter, we will learn about another library, called Valtio, which is a library primarily for module state, with a unique syntax.

9
Use Case Scenario 3 – Valtio

Valtio (`https://github.com/pmndrs/valtio`) is yet another library for global state. Unlike Zustand and Jotai, it's based on the mutating update model. It's primarily for module states like Zustand. It utilizes proxies to get an immutable snapshot, which is required to integrate with React.

The API is just JavaScript and everything works behind the scenes. It also leverages proxies to automatically optimize re-renders. It doesn't require a selector to control re-renders. The automatic render optimization is based on a technique called **state usage tracking**. Using state usage tracking, it can detect which part of the state is used, and it can let a component re-render only if the used part of the state is changed. In the end, developers need to write less code.

In this chapter, we will learn about the basic usage of the Valtio library and how it deals with mutating updates. Snapshots are a key feature to create an immutable state. We will also discuss how snapshots and proxies allow us to optimize re-renders.

In this chapter, we will cover the following topics:

- Exploring Valtio, another module state library
- Utilizing proxies to detect mutations and create an immutable state
- Using proxies to optimize re-renders

- Creating small application code
- The pros and cons of this approach

Technical requirements

You are expected to have moderate knowledge of React, including React Hooks. Refer to the official site, `https://reactjs.org`, to learn more.

In some code, we use TypeScript (`https://www.typescriptlang.org`), and you should have basic knowledge of it.

The code in this chapter is available on GitHub: `https://github.com/PacktPublishing/Micro-State-Management-with-React-Hooks/tree/main/chapter_09`.

To run the code snippets, you need a React environment, for example, Create React App (`https://create-react-app.dev`) or CodeSandbox (`https://codesandbox.io`).

Exploring Valtio, another module state library

Valtio is a library primarily used for module state, which is the same as Zustand.

As we learned in *Chapter 7, Use Case Scenario 1 – Zustand*, we create a store in Zustand as follows:

```
const store = create(() => ({
  count: 0,
  text: "hello",
}));
```

The `store` variable has some properties, one of which is `setState`. With `setState`, we can update the state. For example, the following is incrementing the `count` value:

```
store.setState((prev) => ({
  count: prev.count + 1,
}))
```

Why do we need to use `setState` to update a state value? Because we want to update the state immutably. Internally, the previous `setState` works like the following:

```
moduleState = Object.assign({}, moduleState, {
  count: moduleState.count + 1
});
```

This is the way to update an object immutably.

Let's imagine a case where we don't need to follow the immutable update rule. In this case, the code to increment the `count` value in `moduleState` would be the following:

```
++moduleState.count;
```

Wouldn't it be nice if we could write code like that and make it work with React? Actually, we can implement this with proxies.

A proxy is a special object in JavaScript (`https://developer.mozilla.org/en-US/docs/Web/JavaScript/Reference/Global_Objects/Proxy`). We can define some handlers to trap object operations. For example, you can add a `set` handler to trap object mutations:

```
const proxyObject = new Proxy({
  count: 0,
  text: "hello",
}, {
  set: (target, prop, value) => {
    console.log("start setting", prop);
    target[prop] = value;
    console.log("end setting", prop);
  },
});
```

We create `proxyObject` with new `Proxy` with two arguments. The first argument is an object itself. The second argument is a collection object of handlers. In this case, we have a `set` handler, which traps the `set` operation and adds `console.log` statements.

`proxyObject` is a special object and when you set a value, it will log to the console before and after setting the value. The following is the screen output if you run the code in the Node.js REPL (`https://nodejs.dev/learn/how-to-use-the-nodejs-repl`):

```
> ++proxyObject.count
start setting count
end setting count
1
```

Conceptually, as a proxy can detect any mutations, we could technically use similar behavior to `setState` in Zustand. Valtio is a library that utilizes proxies to detect state mutations.

In this section, we learned that Valtio is a library that uses the mutating update model. Next up, we will learn how Valtio creates immutable states with mutations.

Utilizing proxies to detect mutations and create an immutable state

Valtio creates immutable objects from mutable objects with proxies. We call the immutable object a **snapshot**.

To create a mutable object wrapped in a proxy object, we use the `proxy` function exported by Valtio.

The following example is to create an object with a `count` property:

```
import { proxy } from "valtio";

const state = proxy({ count: 0 });
```

The `state` object returned by the `proxy` function is a proxy object that detects mutations. This allows you to create an immutable object.

To create an immutable object, we use the `snapshot` function exported by Valtio, as follows:

```
import { snapshot } from "valtio";

const snap1 = snapshot(state);
```

Though the `state` variable is `{ count: 0 }` and the `snap1` variable is `{ count: 0 }`, `state` and `snap1` have different references. `state` is a mutable object wrapped in a proxy, whereas `snap1` is an immutable object frozen with `Object.freeze` (`https://developer.mozilla.org/en-US/docs/Web/JavaScript/Reference/Global_Objects/Object/freeze`).

Let's see how snapshots work. We mutate the `state` object and create another snapshot, as follows:

```
++state.count;

const snap2 = snapshot(state);
```

The `state` variable is `{ count: 1 }` and has the same reference as before. The `snap2` variable is `{ count: 1 }` and has a new reference. Because `snap1` and `snap2` are immutable, we can check the equality with `snap1 === snap2`, and know whether anything in the objects differs.

The `proxy` and `snapshot` functions work for nested objects and optimize snapshot creation. That means the `snapshot` function will create a new snapshot only if necessary, that is, when any of its properties are changed. Let's look at another example. `state2` has two nested `c` properties:

```
const state2 = proxy({
  obj1: { c: 0 },
  obj2: { c: 0 },
});

const snap21 = snapshot(state2)

++state2.obj.c;

const snap22 = snapshot(state2)
```

In this case, the `snap21` variable is `{ obj1: { c: 0 }, obj2: { c: 0 } }` and the `snap22` variable is `{ obj1: { c: 1 }, obj2: { c: 0 } }`. `snap21` and `snap22` have difference references, hence `snap21 !== snap22` holds.

How about nested objects? `snap21.obj1` and `snap22.obj1` are different, but `snap21.obj2` and `snap22.obj2` are the same. This is because the value of the internal `c` property of `obj2` isn't changed. `obj2` doesn't need to be changed, hence `snap21.obj2 === snap22.obj2` holds.

This snapshot optimization is an important feature. The fact that `snap21.obj2` and `snap22.obj2` have the same reference means they share memory. Valtio creates snapshots only if necessary, optimizing memory usage. This optimization can be done in Zustand, but it's the developer's responsibility to properly create new immutable states. In contrast, Valtio does the optimization behind the scenes. In Valtio, developers are free from the responsibility of creating new immutable states.

> **Important Note**
> Valtio's optimization is based on caching with a previous snapshot. In other words, the cache size is 1. If we increment the count with `++state.count` and then decrement it with `--state.count`, a new snapshot will be created.

In this section, we learned how Valtio creates immutable state "snapshots" automatically. Next up, we will learn about Valtio's hooks for React.

Using proxies to optimize re-renders

Valtio uses proxies to optimize re-renders, as well as detecting mutations. This is the pattern of optimizing re-renders we learned about in the *Detecting property access* section of *Chapter 6, Introducing Global State Libraries*.

Let's learn about the usage and behavior of Valtio hooks with a counter app. The hook is called `useSnapshot`. The implementation of `useSnapshot` is based on the `snapshot` function and another proxy to wrap it. This `snapshot` proxy has a different purpose from the proxy used in the `proxy` function. The `snapshot` proxy is used to detect the property access of a snapshot object. We will see how render optimization works, thanks to the `snapshot` proxy.

We start with importing functions from Valtio to create a counter app:

```
import { proxy, useSnapshot } from "valtio";
```

`proxy` and `useSnapshot` are two main functions provided by Valtio and they cover most use cases.

We then create a `state` object with `proxy`. In our counter app, there are two counts – `count1` and `count2`:

```
const state = proxy({
  count1: 0,
  count2: 0,
});
```

The `proxy` function takes an initial object and returns a new proxy object. We can mutate the `state` object as we like.

Next, we define the `Counter1` component, which uses the `state` object and shows the `count1` property:

```
const Counter1 = () => {
  const snap = useSnapshot(state);
  const inc = () => ++state.count1;
  return (
    <>
      {snap.count1} <button onClick={inc}>+1</button>
    </>
  );
};
```

It's our convention to set the name of the return value of `useSnapshot` to name. The `inc` action is a function to mutate the `state` object. We mutate the `state` proxy object; snap is only to read. The snap object is frozen with `Object.freeze` (`https://developer.mozilla.org/en-US/docs/Web/JavaScript/Reference/Global_Objects/Object/freeze`) and it can't be mutated technically. Without `Object.freeze`, JavaScript objects are always mutable and we can only treat it as immutable by convention. `snap.count1` is accessing the `count1` property of the `state` object. The access is detected by the `useSnapshot` hook as tracking information, and based on the tracking information, the `useSnapshot` hook triggers re-renders only when necessary.

We define the `Counter2` component likewise:

```
const Counter2 = () => {
  const snap = useSnapshot(state);
  const inc = () => ++state.count2;
  return (
    <>
      {snap.count2} <button onClick={inc}>+1</button>
    </>
  );
};
```

The difference from `Counter1` is it uses the `count2` property instead of the `count1` property. If we want to define a shared component, we can define a single component and take the property name in `props`.

Finally, we define the `App` component. As we don't use Context, there are no providers:

```
const App = () => (
  <>
    <div><Counter1 /></div>
    <div><Counter2 /></div>
  </>
);
```

How does this app work? On the initial render, the `state` object is `{ count1: 0, count2: 0 }` and so is its snapshot object. The `Counter1` component accesses the `count1` property of the snapshot object and the `Counter2` component accesses the `count2` property of the snapshot object. Each `useSnapshot` hook knows and remembers tracking information. The tracking information represents which property is accessed.

When we click the button in the `Counter1` component (the first button in *Figure 9.1*), it increments the `count1` property of the `state` object:

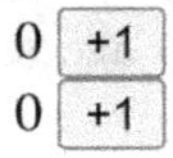

0 [+1]
0 [+1]

Figure 9.1 – First screenshot of the counter app

Thus, the `state` object becomes `{ count1: 1, count2: 0 }`. The `Counter1` component re-renders with the new number 1. However, the `Counter2` component doesn't re-render, because `count2` is still 0 and not changed (*Figure 9.2*):

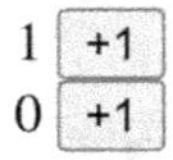

1 [+1]
0 [+1]

Figure 9.2 – Second screenshot of the counter app

Re-renders are optimized with tracking information.

In our counter app, the `state` object is simple with two properties with number values. Valtio supports nested objects and arrays. A contrived example is the following:

```
const contrivedState = proxy({
   num: 123,
   str: "hello",
   arr: [1, 2, 3],
   nestedObject: { foo: "bar" },
   objectArray: [{ a: 1 }, { b: 2 }],
});
```

Basically, any objects containing plain objects and arrays are fully supported even though they are nested deeply. For more information, please refer to the project site: `https://github.com/pmndrs/valtio`.

In this section, we learned how Valtio optimizes re-renders with snapshots and proxies. In the next section, we will learn how to structure an app with an example.

Creating small application code

We will learn how to create a small app. Our example app is a to-do app. Valtio is unopinionated about how to structure apps. This is one of the typical patterns.

Let's look at how a to-do app can be structured. First, we define the `Todo` type:

```
type Todo = {
   id: string;
   title: string;
   done: boolean;
};
```

A `Todo` item has an `id` string value, a `title` string value, and a `done` Boolean value.

We then define a `state` object using the defined `Todo` type:

```
const state = proxy<{ todos: Todo[] }>({
   todos: [],
});
```

The `state` object is created by wrapping an initial object with `proxy`.

To manipulate the `state` object, we define some helper functions – `addTodo` to add a new to-do item, `removeTodo` to remove it, and `toggleTodo` to toggle the done status:

```
const createTodo = (title: string) => {
  state.todos.push({
    id: nanoid(),
    title,
    done: false,
  });
};

const removeTodo = (id: string) => {
  const index = state.todos.findIndex(
    (item) => item.id === id
  );
  state.todos.splice(index, 1);
};

const toggleTodo = (id: string) => {
  const index = state.todos.findIndex(
    (item) => item.id === id
  );
  state.todos[index].done = !state.todos[index].done;
};
```

`nanoid` is a small function to generate a unique ID (https://www.npmjs.com/package/nanoid). Notice these three functions are based on normal JavaScript syntax. They treat `state` just like a normal JavaScript object. This is accomplished with proxies.

The following is the `TodoItem` component, which has a checkbox toggle with the `done` status, text with a different style with the `done` status, and a button to remove the item:

```
const TodoItem = ({
  id,
  title,
  done,
}: {
```

```
      id: string;
      title: string;
      done: boolean;
}) => {
    return (
      <div>
        <input
          type="checkbox"
          checked={done}
          onChange={() => toggleTodo(id)}
        />
        <span
          style={{
            textDecoration: done ? "line-through" : "none",
          }}
        >
          {title}
        </span>
        <button onClick={() => removeTodo(id)}>
          Delete
        </button>
      </div>
    );
};
```

```
const MemoedTodoItem = memo(TodoItem);
```

Notice this component receives the id, title, and done properties separately, instead of receiving the todo object. This is because we use the memo function and create the MemoedTodoItem component. Our state usage tracking detects property access, and if we pass an object to a memoed component, the property access is omitted.

To use the MemoedTodoItem component, the TodoList component is defined with useSnapshot, as follows:

```
const TodoList = () => {
    const { todos } = useSnapshot(state);
    return (
```

```
        <div>
          {todos.map((todo) => (
            <MemoedTodoItem
              key={todo.id}
              id={todo.id}
              title={todo.title}
              done={todo.done}
            />
          ))}
        </div>
      );
  };
```

This component takes `todos` from the result of `useSnapshot` and accesses all properties in objects in the `todos` array. Hence, `useSnapshot` triggers a re-render if any part of `todos` is changed. It's not a big issue and this is a valid pattern because the `MemoedTodoItem` component won't re-render unless `id`, `title`, or `done` is changed. We will learn about another pattern later in this section.

To create a new to-do item, the following is a small component that has a local state for the input field and invokes `createTodo` when the **Add** button is clicked:

```
const NewTodo = () => {
  const [text, setText] = useState("");
  const onClick = () => {
    createTodo(text);
    setText("");
  };
  return (
    <div>
      <input
        value={text}
        onChange={(e) => setText(e.target.value)}
      />
      <button onClick={onClick} disabled={!text}>
        Add
      </button>
    </div>
```

```
  );
};
```

Finally, we combine the defined components in the App component:

```
const App = () => (
  <>
    <TodoList />
    <NewTodo />
  </>
);
```

Let's look at how this app works:

1. At first, it has only a text field and an **Add** button (*Figure 9.3*):

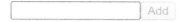

Figure 9.3 – First screenshot of the todos app

2. If we click the **Add** button, a new item is added (*Figure 9.4*):

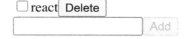

Figure 9.4 – Second screenshot of the todos app

3. We can add as many items as we want (*Figure 9.5*):

Figure 9.5 – Third screenshot of the todos app

4. Clicking a checkbox will toggle the done status (*Figure 9.6*):

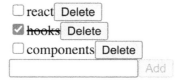

Figure 9.6 – Fourth screenshot of the todos app

5.   Clicking the **Delete** button will delete the item (*Figure 9.7*):

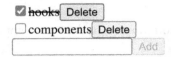

Figure 9.7 – Fifth screenshot of the todos app

The app we created so far works pretty well. But there is room for improvement in terms of extra re-renders. When we toggle the done state of an existing item, not only the corresponding TodoItem component but also the TodoList component will re-render. As noted, this is not a big issue as long as the TodoList component itself is fairly lightweight.

We have another pattern to eliminate the extra re-render in the TodoList component. This doesn't mean the overall performance can always be improved. Which approach we should take depends on the app in question.

In the new approach, we use useSnapshot in each TodoItem component. The TodoItem component only receives the id property. The following is the modified TodoItem component:

```
const TodoItem = ({ id }: { id: string }) => {
  const todoState = state.todos.find(
    (todo) => todo.id === id
  );
  if (!todoState) {
    throw new Error("invalid todo id");
  }
  const { title, done } = useSnapshot(todoState);
  return (
    <div>
      <input
        type="checkbox"
        checked={done}
        onChange={() => toggleTodo(id)}
      />
      <span
        style={{
          textDecoration: done ? "line-through" : "none",
        }}
```

```
      >
        {title}
      </span>
      <button onClick={() => removeTodo(id)}>
        Delete
      </button>
    </div>
  );
};
```

```
const MemoedTodoItem = memo(TodoItem);
```

Based on the id property, it finds todoState, uses useSnapshot with todoState, and gets the title and done properties. This component will re-render only if the id, title, or done properties are changed.

Now, let's look at the modified TodoList component. Unlike the previous one, it only needs to pass the id properties:

```
const TodoList = () => {
  const { todos } = useSnapshot(state);
  const todoIds = todos.map((todo) => todo.id);
  return (
    <div>
      {todoIds.map((todoId) => (
        <MemoedTodoItem key={todoId} id={todoId} />
      ))}
    </div>
  );
};
```

So, todoIds is created from the id property of each todo object. This component will only re-render if the order of id is changed, or if some id is added or removed. If only the done status of an existing item is changed, this component won't re-render. Hence, the extra re-render is eliminated.

In medium-sized apps, the change in the two approaches is subtle in terms of performance. The two approaches are more meaningful for different coding patterns. Developers can choose the one that is more comfortable with their mental model.

In this section, we learned about useSnapshot use cases with a small app. Next up, we will discuss some pros and cons of this library and the approach in general.

The pros and cons of this approach

We have seen how Valtio works and one question is when we should use it and when we should not.

One big aspect is the mental model. We have two state-updating models. One is for immutable updates and the other for mutable updates. While JavaScript itself allows mutable updates, React is built around immutable states. Hence, if we mix the two models, we should be careful not to confuse ourselves. One possible solution would be to clearly separate the Valtio state and React state so that the mental model switch is reasonable. If it works, Valtio can fit in. Otherwise, maybe stick with immutable updates.

The major benefit of mutable updates is we can use native JavaScript functions.

For example, removing an item from an array with an index value can be written as follows:

```
array.splice(index, 1)
```

In immutable updates, this is not so easy. For example, it can be written with slice, as follows:

```
[...array.slice(0, index), ...array.slice(index + 1)]
```

Another example is to change the value in a deeply nested object. It can be done in mutable updates as follows:

```
state.a.b.c.text = "hello";
```

In immutable updates, it has to be something like the following:

```
{
  ...state,
  a: {
    ...state.a,
    b: {
      ...state.a.b,
      c: {
        ...state.a.b.c,
```

```
            text: "hello",
        },
      },
    },
}
```

This is not very pleasant to write. Valtio helps to reduce application code with mutable updates.

Valtio also helps to reduce application code with proxy-based render optimization.

Suppose we have a state with the `count` and `text` properties, as follows:

```
const state = proxy({ count: 0, text: "hello" });
```

If we use only `count` in a component, we can write the following in Valtio:

```
const Component = () => {
  const { count } = useSnapshot(state);
  return <>{count}</>;
};
```

In comparison, with Zustand, this will be something like the following:

```
const Component = () => {
  const count = useStore((state) => state.count);
  return <>{count}</>;
};
```

The difference is trivial, but we have `count` in two places.

Let's look at a contrived scenario. Suppose we want to show the `text` value if the `showText` property is truthy. With `useSnapshot`, it can be done as follows:

```
const Component = ({ showText }) => {
  const snap = useSnapshot(state);
  return <>{snap.count} {showText ? snap.text : ""}</>;
};
```

Implementing the same behavior with selector-based hooks is tough. One solution is to use a hook twice. With Zustand, it will be like the following:

```
const Component = ({ showText }) => {
  const count = useStore((state) => state.count);
  const text = useStore(
    (state) => showText ? state.text : ""
  );
  return <>{count} {text}</>;
};
```

This means if we have more conditions, we need more hooks.

On the other hand, a disadvantage of proxy-based render optimization can be less predictability. Proxies take care of render optimization behind the scenes and sometimes it's hard to debug the behavior. Some may prefer explicit selector-based hooks.

In summary, there's no one-size-fits-all solution. It's up to developers to choose the solution that fits their needs.

In this section, we discussed the approach taken in the Valtio library.

Summary

In this chapter, we learned about a library called Valtio. It utilizes proxies extensively. We've seen examples and learned how it can be used. It allows mutating state, which feels like using normal JavaScript objects, and the proxy-based render optimization helps reduce application code. It depends on developers' requirements whether this approach is a good choice.

In the next chapter, we will learn about another library, called React Tracked, which is a library that is based on Context and has proxy-based render optimization like Valtio.

10
Use Case Scenario 4 – React Tracked

React Tracked (`https://react-tracked.js.org`) is a library for state usage tracking that optimizes re-renders automatically based on property access. It provides the same functionality to eliminate extra re-renders as Valtio, which we discussed in *Chapter 9, Use Case Scenario 3 – Valtio.*

React Tracked can be used with other state management libraries. The primary use case is `useState` or `useReducer`, but it can also be used with Redux (`https://redux.js.org`), Zustand (discussed in *Chapter 7, Use Case Scenario 1 – Zustand*), and other similar libraries.

In this chapter, we will again discuss optimizing re-renders with state usage tracking and compare related libraries. We will learn two usages of React Tracked, one with `useState` and the other with React Redux (`https://react-redux.js.org`). We will wrap up with a look at how React Tracked will work with the future version of React.

In this chapter, we will cover the following topics:

- Understanding React Tracked
- Using React Tracked with `useState` and `useReducer`

- Using React Tracked with React Redux
- Future prospects

Technical requirements

You are expected to have a moderate amount of knowledge about React, including React Hooks. Refer to the official site, `https://reactjs.org`, to learn more.

In some code, we use TypeScript (`https://www.typescriptlang.org`), and you should have basic knowledge of it.

The code in this chapter is available on GitHub: `https://github.com/PacktPublishing/Micro-State-Management-with-React-Hooks/tree/main/chapter_10`.

To run the code snippets, you need a React environment – for example, Create React App (`https://create-react-app.dev`) or CodeSandbox (`https://codesandbox.io`).

Understanding React Tracked

We have been learning about several global state libraries, but React Tracked is slightly different from the ones we have learned about so far. React Tracked doesn't provide state functionality, but what it does provide is render optimization functionality. We call this functionality **state usage tracking**.

Let's recap how React Context behaves because one of the use cases of state usage tracking in React Tracked is for a React Context.

Suppose we define a Context with `createContext` as follows:

```
const NameContext = createContext([
  { firstName: 'react', lastName: 'hooks' },
  () => {},
]);
```

`createContext` takes an initial value, which is an array in this case. The first item in the array is an initial state object. The second item in the array, `() => {}`, is a dummy updating function.

The reason we put such an array as the initial value is to match the return value of `useState`. We often define `NameProvider` with `useState` for a global state:

```
const NameProvider = ({ children }) => (
  <NameContext.Provider
    value={
      useState({ firstName: 'react', lastName: 'hooks' })
    }
  >
    {children}
  </NameContext.Provider>
};
```

You should usually use the `NameProvider` component in a root component or some component close to it.

Now that we have the `NameProvider` component, we can consume it under its tree. To consume the Context value, we use `useContext`. Let's assume we only need `firstName` and define a `useFirstName` hook:

```
const useFirstName = () => {
  const [{ firstName }] = useContext(NameContext);
  return firstName;
};
```

This works fine. However, there's a possibility of extra re-renders. If we update only `lastName` without changing `firstName`, the new Context value will be propagated and `useContext(NameContext)` triggers a re-render. The `useFirstName` hook only reads `firstName` from the Context value. Hence, this becomes an extra re-render.

This behavior is obvious from an implementation point of view. But from a developer's point of view, it doesn't seem ideal because it only uses `firstName` from the Context value. From the developer's point of view, the expectation would be that it doesn't depend on other properties – in this case, `lastName`.

State usage tracking is the feature that realizes this expected behavior. If we only use `firstName` in the state object, we expect the hook to trigger re-renders only when `firstName` changes. This can be accomplished with proxies.

React Tracked allows us to define a hook called `useTracked`, which can be used instead of `useContext(NameContext)`. `useTracked` wraps the state with proxies and tracks its usage. The expected usage of `useTracked` looks like the following:

```
const useFirstName = () => {
  const [{ firstName }] = useTracked();
  return firstName;
};
```

The usage doesn't differ from the usage of `useContext(NameContext)`. This is the whole point of state usage tracking. Our code looks just as usual, but behind the scenes, it tracks the state usage and optimizes renders automatically.

Automatic render optimization was discussed in *Chapter 9, Use Case Scenario 3 – Valtio*. React Tracked and Valtio use the same state usage tracking feature. Actually, they use the same internal library, which is called `proxy-compare`: `https://github.com/dai-shi/proxy-compare`.

In this section, we revisited state usage tracking and learned how it can optimize re-renders. In the next section, we will learn how to use React Tracked with `useState` and `useReducer`.

Using React Tracked with useState and useReducer

The primary use case of React Tracked is to replace a use case of React Context. The API in React Tracked is specifically designed for this use case.

We will explore two usages with `useState` and `useReducer`. First, let's learn about the usage with `useState`.

Using React Tracked with useState

Before exploring the usage of React Tracked with `useState`, let's revisit how we can create a global state with React Context.

We first create a custom hook, which calls `useState` with an initial state value:

```
const useValue = () =>
  useState({ count: 0, text: "hello" });
```

Defining the custom hook is good for TypeScript because you can grab the type with the `typeof` operator.

The following is a definition of our Context:

```
const StateContext = createContext<
  ReturnType<typeof useValue> | null
>(null);
```

It has a type annotation in TypeScript. The default value is `null`.

To use the Context, we need a `Provider` component. The following is a custom `Provider` that uses `useValue` for the Context value:

```
const Provider = ({ children }: { children: ReactNode }) => (
  <StateContext.Provider value={useValue()}>
    {children}
  </StateContext.Provider>
);
```

This is a component that injects the `StateContext.Provider` component. As we defined `useValue` separately, the implementation of `Provider` can use it in **JavaScript Syntax Extension (JSX)**.

To consume the Context's value, we use `useContext`. We define a custom hook as follows:

```
const useStateContext = () => {
  const contextValue = useContext(StateContext);
  if (contextValue === null) {
    throw new Error("Please use Provider");
  }
  return contextValue;
};
```

This custom hook checks the existence of `Provider` by comparing `contextValue` with `null`. If it's `null`, it throws an error, and developers will notice that `Provider` is missing.

Now, it's time to define some components for the app. The first component is `Counter`, which shows the `count` property of the state as well as a button to increment the count value:

```
const Counter = () => {
  const [state, setState] = useStateContext();
  const inc = () => {
    setState((prev) => ({
      ...prev,
      count: prev.count + 1,
    }));
  };
  return (
    <div>
      count: {state.count}
      <button onClick={inc}>+1</button>
    </div>
  );
};
```

Note that `useStateContext` returns a tuple of the `state` value and the updating function. This is exactly the same as what `useValue` returns.

Next, we define the second component, `TextBox`, which shows an input field for the `text` property of the state:

```
const TextBox = () => {
  const [state, setState] = useStateContext();
  const setText = (text: string) => {
    setState((prev) => ({ ...prev, text }));
  };
  return (
    <div>
      <input
        value={state.text}
        onChange={(e) => setText(e.target.value)}
      />
    </div>
  );
};
```

We again use `useStateContext` and get the `state` value and the `setState` function. The `setText` function takes a string argument and invokes the `setState` function.

Finally, we define the `App` component, which has the `Provider`, `Counter`, and `TextBox` components in it:

```
const App = () => (
  <Provider>
    <div>
      <Counter />
      <Counter />
      <TextBox />
      <TextBox />
    </div>
  </Provider>
);
```

How does this app behave? The Context handles the state object as a whole, and `useContext` will trigger re-renders when the state object changes. Even if only a single property changes in the state object, all `useContext` hooks trigger re-renders. This means that if we click a button in the `Counter` component, it increments the `count` property of the state object, and it causes both the `Counter` and `TextBox` components to re-render. While the `Counter` component re-renders with the new `count` value, the `TextBox` component re-renders with the same `text` value. This is an extra re-render.

The extra re-render behavior with Context is expected, and if we want to avoid it, we should split it into smaller pieces. Refer to *Chapter 3*, *Sharing the Component State with Context*, to learn more about best practices with React Context.

Now, what does it look like with React Tracked? Let's convert the previous example to a new example with React Tracked. First, we import `createContainer` from the React Tracked library:

```
import { createContainer } from "react-tracked";
```

We then use the `useValue` hook defined in `const useValue = () =>` `useState({ count: 0, text: "hello" });` and call the `createContainer` function:

```
const { Provider, useTracked } =
  createContainer(useValue);
```

From the results, `Provider` and `useTracked` are extracted. The `Provider` component can be used in the same way as in the previous example of this section. The `useTracked` hook can be used in the same way as the `useStateContext` hook we defined in the previous example of this section.

Using the new `useTracked` hook, the `Counter` component becomes as follows:

```
const Counter = () => {
  const [state, setState] = useTracked();
  const inc = () => {
    setState(
      (prev) => ({ ...prev, count: prev.count + 1 })
    );
  };
  return (
    <div>
      count: {state.count}
      <button onClick={inc}>+1</button>
    </div>
  );
};
```

We simply replaced `useStateContext` with `useTracked`. The rest of the code is the same.

Likewise, the following is the new `TextBox` component:

```
const TextBox = () => {
  const [state, setState] = useTracked();
  const setText = (text: string) => {
    setState((prev) => ({ ...prev, text }));
  };
  return (
    <div>
      <input
        value={state.text}
        onChange={(e) => setText(e.target.value)}
      />
    </div>
```

```
    );
  };
```

The only change is the replacement of `useStateContext` with `useTracked`.

The `App` component is exactly the same as in the previous example of this section, using the new `Provider` component:

```
const App = () => (
  <Provider>
    <div>
      <Counter />
      <Counter />
      <TextBox />
      <TextBox />
    </div>
  </Provider>
);
```

How does this new app behave? The `state` object returned by `useTracked` is tracked, which means the `useTracked` hook remembers which properties of `state` are accessed. The `useTracked` hook will trigger a re-render only if the accessed properties are changed. Hence, if you click a button in the `Counter` component, only the `Counter` component re-renders, and the `TextBox` component doesn't re-render, as shown here:

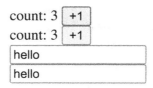

Figure 10.1 – A screenshot of the app with React Tracked and useState

Essentially, what we changed is `createContainer` instead of `createContext`, and `useTracked` instead of `useStateContext`. The result gives us optimized re-renders. This is the state usage tracking feature.

The `useValue` custom hook we passed to the `createContainer` function can be anything as long as it returns a tuple such as `useState`. Let's look at another example using `useReducer`.

Using React Tracked with useReducer

In this example, we use useReducer instead of useState. The useReducer hook is an advanced hook with more features, but it's mostly syntactic difference. Refer to the *Exploring the similarity and difference between useState and useReducer* section in *Chapter 1, What Is Micro State Management with React Hooks?*, for more detailed discussions.

> **Important Note about useReducer**
>
> The useReducer hook is an official React hook. It takes a reducer function to update states. A reducer function is a programming pattern, not related to React or even JavaScript. The useReducer hook applies the pattern to states. The reducer function in React is popularized by Redux. The useReducer covers Redux's use cases in terms of the reduce pattern. However, it doesn't cover the other Redux use cases, such as React Redux and store enhancer or middleware. The useReducer hook accepts any kind of actions unlike Redux, which requires an action to be an object with a type property.

The new useValue hook uses useReducer and useEffect. useReducer is defined with a reducer function and an initial state. useEffect has a function that logs the state value to the console. The following is the useValue code in TypeScript:

```typescript
const useValue = () => {
  type State = { count: number; text: string };
  type Action =
    | { type: "INC" }
    | { type: "SET_TEXT"; text: string };
  const [state, dispatch] = useReducer(
    (state: State, action: Action) => {
      if (action.type === "INC") {
        return { ...state, count: state.count + 1 };
      }
      if (action.type === "SET_TEXT") {
        return { ...state, text: action.text };
      }
      throw new Error("unknown action type");
    },
    { count: 0, text: "hello" }
  );
  useEffect(() => {
```

```
    console.log("latest state", state);
  }, [state]);
  return [state, dispatch] as const;
};
```

The reducer function accepts action types of INC and SET_TEXT. The useEffect hook is used in console logging, but it's not limited to it. For example, it can interact with remote resources. The useValue hook returns a tuple of state and dispatch. As long as the return tuple follows this shape, we can implement the hook as we like. For example, we could use more than one useState hook.

Using the new useValue hook, we run createContainer:

```
const { Provider, useTracked } = createContainer(useValue);
```

The way we use createContainer doesn't change, even if we change useValue.

Using the new useTracked hook, we implement the Counter component:

```
const Counter = () => {
  const [state, dispatch] = useTracked();
  const inc = () => dispatch({ type: "INC" });
  return (
    <div>
      count: {state.count}
      <button onClick={inc}>+1</button>
    </div>
  );
};
```

Because useTracked returns the same shaped tuple as useValue returns, we name the second item in the tuple dispatch, which is a function that dispatches an action. The Counter component dispatches an INC action.

Next is the TextBox component:

```
const TextBox = () => {
  const [state, dispatch] = useTracked();
  const setText = (text: string) => {
    dispatch({ type: "SET_TEXT", text });
  };
```

```
    return (
      <div>
        <input
          value={state.text}
          onChange={ (e) => setText(e.target.value) }
        />
      </div>
    );
};
```

Likewise, the dispatch function is used for a SET_TEXT action.

Finally, we have the App component:

```
const App = () => (
  <Provider>
    <div>
      <Counter />
      <Counter />
      <TextBox />
      <TextBox />
    </div>
  </Provider>
);
```

The behavior of the new App component is exactly the same as the previous one. The difference between the examples with useState and useReducer is that useValue returns a tuple of state and dispatch; thus useTracked also returns a tuple of state and dispatch.

The reason why React Tracked can optimize re-renders is not only state usage tracking but also its internal library called use-context-selector (https://github.com/dai-shi/use-context-selector). It allows us to subscribe to the Context value with a selector function. This subscription bypasses the limitations of React Context.

In this section, we saw a basic example with bare React Context, and two examples with React Tracked with useState and useReducer. In the next section, we will learn a usage of React Tracked with React Redux, which uses the state usage tracking feature without use-context-selector.

Using React Tracked with React Redux

The primary use case of React Tracked is to replace a use case of React Context. This is done by using `use-context-selector` internally.

React Tracked exposes a low-level function called `createTrackedSelector` to cover non-React Context use cases. It takes a hook called `useSelector` and returns a hook called `useTrackedState`:

```
const useTrackedState = createTrackedSelector(useSelector);
```

`useSelector` is a hook that takes a selector function and returns the result of the selector function. It will trigger re-renders when the result changes. `useTrackedState` is a hook that returns an entire state wrapped in proxies to track the `state` usage.

Let's look at a concrete example with React Redux. This provides a `useSelector` hook, and it's straightforward to apply `createTrackedSelector`.

> **Important Note about React Redux**
>
> React Redux uses React Context internally, but it doesn't use Context for propagating a state value. It uses React Context for dependency injection, and the state propagation is done by subscription. React Redux's `useSelector` is optimized to re-render only if the selector result changes. This is not possible with Context propagation at the time of writing. There are many other libraries that take the same approach, and in fact, the `use-context-selector` UserLand solution is the same too.

First, we import some functions from libraries, namely `redux`, `react-redux`, and `react-tracked`:

```
import { createStore } from "redux";
import {
  Provider,
  useDispatch,
  useSelector,
} from "react-redux";
import { createTrackedSelector } from "react-tracked";
```

The first two import lines are a traditional React Redux setup. The third line is our addition.

Next, we define a Redux store with `initialState` and `reducer`:

```
type State = { count: number; text: string };
type Action =
  | { type: "INC" }
  | { type: "SET_TEXT"; text: string };

const initialState: State = { count: 0, text: "hello" };
const reducer = (state = initialState, action: Action) => {
  if (action.type === "INC") {
    return { ...state, count: state.count + 1 };
  }
  if (action.type === "SET_TEXT") {
    return { ...state, text: action.text };
  }
  return state;
};


const store = createStore(reducer);
```

This is one traditional way to create a Redux store. Note that it has nothing to do with React Tracked, and any way of creating a Redux store would work.

`createTrackedSelector` allows us to create the `useTrackedState` hook from the `useSelector` hook, which is imported directly from `react-redux`:

```
const useTrackedState =
  createTrackedSelector<State>(useSelector);
```

We need to explicitly type the hook with `<State>`.

Using `useTrackedState`, the `Counter` component is defined as follows:

```
const Counter = () => {
  const dispatch = useDispatch();
  const { count } = useTrackedState();
  const inc = () => dispatch({ type: "INC" });
  return (
```

```
      <div>
        count: {count} <button onClick={inc}>+1</button>
      </div>
    );
  };
```

This should be mostly like a normal React Redux pattern except for the useTrackedState line. In React Redux, it would be as follows:

```
    const count = useSelector((state) => state.count);
```

The change may seem trivial, but with useSelector, developers have more control and responsibility for re-renders, whereas with useTrackedState, the hook controls re-renders automatically.

Likewise, the TextBox component is implemented as follows:

```
  const TextBox = () => {
    const dispatch = useDispatch();
    const state = useTrackedState();
    const setText = (text: string) => {
      dispatch({ type: "SET_TEXT", text });
    };
    return (
      <div>
        <input
          value={state.text}
          onChange={(e) => setText(e.target.value)}
        />
      </div>
    );
  };
```

Again, we used `useTrackedState` instead of `useSelector` for automatic render optimization. To explain how automatic render optimization is useful, let's imagine that `TextBox` takes a `showCount` prop property, which is a Boolean value to show the `count` value in `state`. We can modify the `TextBox` component as follows:

```
const TextBox = ({ showCount }: { showCount: boolean }) => {
  const dispatch = useDispatch();
  const state = useTrackedState();
  const setText = (text: string) => {
    dispatch({ type: "SET_TEXT", text });
  };
  return (
    <div>
      <input
        value={state.text}
        onChange={(e) => setText(e.target.value)}
      />
      {showCount && <span>{state.count}</span>}
    </div>
  );
};
```

Note that we didn't change the `useTrackedState` line at all. With a single `useSelector`, implementing the same behavior would be difficult.

Finally, the following is the `App` component to combine all components:

```
const App = () => (
  <Provider store={store}>
    <div>
      <Counter />
      <Counter />
      <TextBox />
      <TextBox />
    </div>
  </Provider>
);
```

This is exactly the same as using normal React Redux without React Tracked. The re-renders are optimized in this app, which means clicking a button only triggers the `Counter` component to re-render, and the `TextBox` component won't re-render, as shown in the following figure:

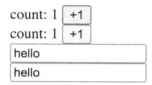

Figure 10.2 – A screenshot of the app with React Tracked and React Redux

In this section, we learned how to use React Tracked with a non-React Context use case. Next, we will discuss what React Tracked can look like with a future version of React.

Future prospects

The implementation of React Tracked depends on two internal libraries:

- `proxy-compare` (https://github.com/dai-shi/proxy-compare)
- `use-context-selector` (https://github.com/dai-shi/use-context-selector)

As we learned in the *Using React Tracked with useState and useReducer* section and the *Using React Tracked with React Redux* section, there are two ways to use React Tracked. The first way is via React Context with `createContainer` and the second is via React Redux with `createTrackedSelector`. The base function is `createTrackedSelector`, which is implemented with the `proxy-compare` library. The `createContainer` function is a higher abstraction, which is implemented with `createTrackedSelector` and the `use-context-selector` library.

In terms of the use of Context in React Tracked, the `use-context-selector` library is important. What is the role of `use-context-selector`? It provides a `useContextSelector` hook. As we learned in the *Understanding Context* section in *Chapter 3, Sharing the Component State with Context*, React Context is designed so that all Context consumer components re-render when the Context value is changed. There is a proposal to improve the Context behavior – `useContextSelector`. The `use-context-selector` library is a Userland library that emulates the proposed `useContextSelector` hook as much as possible.

It's very uncertain at the point of writing, but a future version of React may implement `useContextSelector`, or a similar form of it. In this situation, React Tracked can easily migrate from the `use-context-selector` library to a native `useContextSelector`. Hopefully, this should give full compatibility with React features.

Abstracting `use-context-selector` away in the implementation of React Tracked helps migration. If React has an official `useContextSelector` hook in the future, React Tracked can migrate without changing its public API. In this implementation design, `createTrackedSelector` is a building block function in React Tracked, and `createContainer` is a glue function. Exporting both functions allows us to have both usages.

In this section, we discussed the implementation design of React Tracked and how it can migrate to a possible future version of React.

Summary

In this chapter, we learned about a library – React Tracked. This library has two purposes. One purpose is to replace the use case of React Context. The other purpose is to enhance the selector hook provided by some other libraries, such as React Redux.

Technically, the React Tracked library is not a global state library. It's to be used with state functions, such as `useState` and `useReducer`, or Redux. All React Tracked provides is a feature to optimize re-renders.

In the next chapter, we will compare the three libraries for global state, namely Zustand, Jotai, and Valtio, and discuss global state patterns to wrap up this book.

11
Similarities and Differences between Three Global State Libraries

In this book, we introduced three global state libraries: Zustand, Jotai, and Valtio. Let's discuss some similarities and differences between them. These three libraries have some comparable features.

Zustand is similar to Redux (and React Redux) in terms of usage and the store model but, unlike Redux, it's not based on reducers.

Jotai is similar to Recoil (https://recoiljs.org) in terms of the API, but its goal is more to provide a minimal API for non-selector-based render optimization.

Valtio is similar to MobX in terms of the mutating update model, but the level of similarity is only minor, and the render optimization implementation is very different.

All three libraries provide primitive features that fit with micro-state management. They differ in their coding style and approach to render optimization.

In this chapter, we discuss each library by pairing it with its comparable library and then discuss the similarities and differences between the three. We will cover the following topics:

- Differences between Zustand and Redux
- Understanding when to use Jotai and Recoil
- Using Valtio and MobX
- Comparing Zustand, Jotai, and Valtio

Technical requirements

You are expected to have a moderate knowledge of React, including React hooks. Refer to the official site, `https://reactjs.org`, to learn more.

In some code, we use TypeScript (`https://www.typescriptlang.org`), and you should have a basic knowledge of it.

The code in this chapter is available on GitHub: `https://github.com/PacktPublishing/Micro-State-Management-with-React-Hooks/tree/main/chapter_11`.

To run the code snippets, you need a React environment, for example, Create React App (`https://create-react-app.dev`) or CodeSandbox (`https://codesandbox.io`).

Differences between Zustand and Redux

In some use cases, the developer experience can be similar in Zustand and Redux. Both are based on one-way data flow. In one-way data flow, we dispatch `action`, which represents a command to update a state, and after the state is updated with `action`, the new state is propagated to where it's needed. This separation of dispatching and propagating simplifies the flow of data and makes the entire system more predictable.

On the other hand, they differ in how to update states. Redux is based on reducers. A reducer is a pure function that takes a previous state and an `action` object and returns a new state. While updating states with reducers is a strict method, it leads to more predictability. Zustand takes a flexible approach and it doesn't necessarily use reducers to update states.

In this section, we will see a comparison by converting an example with Redux into Zustand. Then we will see the differences between the two.

Example with Redux and Zustand

Let's look at one of the official Redux tutorials. This is the so-called modern Redux with the Redux toolkit: `https://redux-toolkit.js.org/tutorials/quick-start`.

To create a Redux store, we can use `configureStore` from the Redux Toolkit library:

```
// src/app/store.js
import { configureStore } from "@reduxjs/toolkit";
import counterReducer from "../features/counter/counterSlice";

export const store = configureStore({
  reducer: {
    counter: counterReducer,
  },
});
```

The `configureStore` function takes reducers and returns a `store` variable. In this case, it uses one reducer – `counterReducer`.

`counterReducer` is defined in a separate file, using `createSlice` from the Redux Toolkit library. First, we import `createSlice` and define `initialState`:

```
// features/counter/counterSlice.js
import { createSlice } from "@reduxjs/toolkit";

const initialState = {
  value: 0,
};
```

We then define `counterSlice` using `createSlice` and `initialState`:

```
export const counterSlice = createSlice({
  name: "counter",
  initialState,
  reducers: {
    increment: (state) => {
      state.value += 1;
    },
    decrement: (state) => {
```

```
      state.value -= 1;
    },
    incrementByAmount: (
      state,
      action: PayloadAction<number>
    ) => {
      state.value += action.payload;
    },
  },
});
```

The `counterSlice` variable created with the `createSlice` function contains both a reducer and actions. To make them easily importable, we extract the reducer and action properties and export them separately:

```
export const {
  increment,
  decrement,
  incrementByAmount
} = counterSlice.actions;
export default counterSlice.reducer;
```

Next is the `Counter` component, which uses the created store. First, we import two hooks from the `react-redux` library and two actions from the `counterSlice` file:

```
// features/counter/Counter.jsx
import { useSelector, useDispatch } from "react-redux";
import { decrement, increment } from "./counterSlice";
```

We then define the `Counter` component:

```
export function Counter() {
  const count = useSelector((
    state: { counter: { value: number; }; }
  ) => state.counter.value);
  const dispatch = useDispatch();
  return (
    <div>
```

```
          <button onClick={() => dispatch(increment())}>
            Increment
          </button>
          <span>{count}</span>
          <button onClick={() => dispatch(decrement())}>
            Decrement
          </button>
      </div>
    );
}
```

This component uses useSelector and useDispatch hooks from the React Redux library. We use a selector function to get the count value from the store state. Notice that this component doesn't use the created store directly. The useSelector hook takes the store from Context.

Finally, the App component looks like the following:

```
// App.jsx
import { Provider } from "react-redux";
import { store } from "./app/store";
import { Counter } from "./features/counter/Counter";

const App = () => (
    <Provider store={store}>
      <div>
        <Counter />
        <Counter />
      </div>
    </Provider>
);


export default App;
```

We pass the store variable we created with the Provider component. This allows the useSelector hook in the Counter component to access the store variable.

As shown in *Figure 11.1*, this works as expected. We have two `Counter` components in the `App` component, and they share the same `count` value.

Figure 11.1 – Screenshot of the app with Redux

Now, let's see how this can be implemented in Zustand.

First, we create a store with the `create` function from the Zustand library. We begin with importing the Zustand library:

```
// store.js
import create from "zustand";
```

We then define `State` type for TypeScript:

```
type State = {
  counter: {
    value: number;
  };
  counterActions: {
    increment: () => void;
    decrement: () => void;
    incrementByAmount: (amount: number) => void;
  };
};
```

The following is a `store` definition. In Zustand, a hook `useStore` represents a `store`:

```
export const useStore = create<State>((set) => ({
  counter: { value: 0 },
  counterActions: {
    increment: () =>
      set((state) => ({
        counter: { value: state.counter.value + 1 },
      })),
```

```
      decrement: () =>
        set((state) => ({
          counter: { value: state.counter.value - 1 },
        })),
      incrementByAmount: (amount: number) =>
        set((state) => ({
          counter: { value: state.counter.value + amount },
        })),
    },
  }));
```

This defines both the counter state and counter actions in the `store`. The reducer logic is implemented in the function body of the actions.

Next is the `Counter` component, which uses the created store:

```
// Counter.jsx
import { useStore } from "./store";

export function Counter() {
  const count = useStore((state) => state.counter.value);
  const { increment, decrement } = useStore(
    (state) => state.counterActions
  );

  return (
    <div>
      <div>
        <button onClick={increment}>Increment</button>
        <span>{count}</span>
        <button onClick={decrement}>Decrement</button>
      </div>
    </div>
  );
}
```

We use the `useStore` hook to get the `count` value and the actions to update the `count` value. Notice that the `useStore` hook is directly imported from the store file.

Finally, the `App` component looks like the following:

```
// App.jsx
import { Counter } from "./Counter";

const App = () => (
  <div>
    <Counter />
    <Counter />
  </div>
);


export default App;
```

As we don't use Context, we don't need a provider component.

Now, let's discuss a comparison of the two.

Comparing examples of Redux and Zustand

While two implementations of the example in the *Example with Redux and Zustand* section share some common concepts, there are notable differences:

- One of the biggest differences between the example of Redux and Zustand is the directory structure. Modern Redux suggests the `features` directory structure and the `createSlice` function is designed to follow the feature directory pattern. This is a useful pattern for large-scale apps. Zustand, on the other hand, is unopinionated regarding the structure. It's up to developers how to organize files and directories. While it's possible to follow the `features` directory structure with Zustand, there's no specific support from the library. Our Zustand example shows a pattern with `counterActions`, but it's only one possible pattern.

- Another difference in the store creation code is the use of Immer (`https://immerjs.github.io/immer/`). Immer allows a mutation style such as `state.value += 1;`. Modern Redux uses Immer by default. Zustand doesn't use it by default, and neither does our example. It is possible to use Immer in Zustand, but it's optional.

- In terms of store propagation, Redux uses Context, whereas Zustand uses module imports. Context allows the store to be injected at runtime, which works better in some use cases. Zustand optionally supports Context usage.

- Most importantly, Redux Toolkit is based on Redux, which is based on a one-way data flow. So, updating the state in Redux requires actions to be dispatched. This limitation is sometimes good for maintainability and scalability. Zustand is unopinionated regarding the data flow, and while it can be used for one-way data flow, there is no library support and developers need to take care of everything.

In summary, modern Redux is more opinionated about how to manage the state and Zustand is less opinionated about it. In the end, Zustand is a minimalistic library, while Redux and its family are a set of full-featured libraries. The usages of both modern Redux and Zustand seem similar, but the philosophies behind them are different.

In this section, we saw a comparison between modern Redux and Zustand. Next up, we will compare Recoil and Jotai.

Understanding when to use Jotai and Recoil

Jotai's API is highly inspired by Recoil. In the beginning, it's intentionally designed to help migration from Recoil to Jotai. In this section, we will see a comparison by converting an example with Recoil into Jotai. Then, we will discuss the differences between the two.

Example with Recoil and Jotai

Let's look at the Recoil tutorial at `https://recoiljs.org/docs/introduction/getting-started` and see how an example in the Recoil tutorial is converted to Jotai.

To start with the Recoil example, we need to import some functions from the Recoil library:

```
import {
  RecoilRoot,
  atom,
  selector,
  useRecoilState,
  useRecoilValue,
} from "recoil";
```

There are five of them used in this example.

The first state for the text string is created with the `atom` function:

```
const textState = atom({
  key: "textState",
  default: "",
});
```

It takes two properties – the `key` string and the `default` value.

To use the defined state, we use the `useRecoilState` hook:

```
const TextInput = () => {
  const [text, setText] = useRecoilState(textState);
  return (
    <div>
      <input
        type="text"
        value={text}
        onChange={(event) => {
          setText(event.target.value);
        }}
      />
      <br />
      Echo: {text}
    </div>
  );
};
```

`useRecoilState` returns the same value as `useState`. Hence, the rest of the code should be familiar.

The second state is a derived state. We use the `selector` function to define a derived state:

```
const charCountState = selector({
  key: "charCountState",
  get: ({ get }) => get(textState).length,
});
```

It takes two properties – a `key` string and a `get` function. The `get` property is a function that returns a derived value. Another `get` function within a `get` property returns the value of other states created by other `atom` and `selector` functions.

To use the second state, we use the `useRecoilValue` hook, which returns only the value part of the state:

```
const CharacterCount = () => {
  const count = useRecoilValue(charCountState);
  return <>Character Count: {count}</>;
};
```

This component will re-render when `textState` changes because `charCountState` is derived from it.

The `CharacterCounter` component is defined as follows to combine two components that are already defined:

```
const CharacterCounter = () => (
  <div>
    <TextInput />
    <CharacterCount />
  </div>
);
```

Finally, we define the `App` component:

```
const App = () => (
  <RecoilRoot>
    <CharacterCounter />
  </RecoilRoot>
);
```

In the `App` component, we use the `RecoilRoot` component, which holds state values.

As shown in *Figure 11.2*, this app works like this: if you type something in the text field, the text will be shown below the text field, and also the number of characters is shown as follows:

```
hello
```
Echo: hello
Character Count: 5

Figure 11.2 – Screenshot of the app with Recoil

Now, let's convert this example code into Jotai.

We first import two functions from the Jotai library:

```
import { atom, useAtom } from "jotai";
```

Jotai's API tries to be minimal, and the minimal usage requires two functions.

The first atom for the text string is created with the `atom` function:

```
const textAtom = atom("");
```

This is almost the same as Recoil, except that it only has the `default` value because Jotai doesn't require the `key` string. Suffixing the variable name with `Atom` instead of `State` is a convention that is technically unimportant.

To use the defined atom, we use the `useAtom` function:

```
const TextInput = () => {
  const [text, setText] = useAtom(textAtom);
  return (
    <div>
      <input
        type="text"
        value={text}
        onChange={(event) => {
          setText(event.target.value);
        }}
      />
```

```
      <br />
      Echo: {text}
    </div>
  );
};
```

The useAtom function works like useState, and the rest of the code should be familiar to people who are used to useState.

The second atom is a derived atom, which is defined with the atom function:

```
const charCountAtom = atom((get) => get(textAtom).length);
```

In this case, we pass a function to the atom function. The internal function computes the derived value.

To use the second atom, we again use the useAtom function:

```
const CharacterCount = () => {
  const [count] = useAtom(charCountAtom);
  return <>Character Count: {count}</>;
};
```

It's required to get the first part of the returned value with [count]. Other than that, the code and the behavior should be similar to Recoil.

The CharacterCounter component is defined as follows to combine two components that are already defined:

```
const CharacterCounter = () => (
  <div>
    <TextInput />
    <CharacterCount />
  </div>
);
```

Finally, we define the `App` component:

```
const App = () => (
  <>
    <CharacterCounter />
  </>
);
```

The minimal use case of Jotai doesn't require a `Provider` component.

The conversion from the Recoil example to the Jotai example is mostly syntactic, and the behavior is the same.

Let's discuss some of the differences.

Comparing examples of Recoil and Jotai

Although there are many differences in terms of features we didn't use in the example, we'll keep our discussion within the scope of the example we showed, as follows:

- The biggest difference is the existence of the `key` string. One of the big motivations of developing Jotai is to omit the `key` string. Thanks to this feature, the `atom` (`{ key: "textState", default: "" }`) atom definition in Recoil can be `atom("")` in Jotai. Technically, it looks straightforward, but this makes a huge difference to the developer experience. Naming is a hard task in coding, especially because the `key` property has to be unique. Implementation-wise, Jotai utilizes `WeakMap` and relies on the reference of atom objects. On the other hand, Recoil is based on the `key` strings, which don't rely on object references. The benefit of `key` strings is that they're serializable. This should facilitate implementing persistence, which requires serialization. Jotai would require some techniques to overcome serialization.

- Another difference related to the `key` string is the unified `atom` function. The `atom` function in Jotai works for both `atom` and `selector` in Recoil. However, there's a downside. It can't be fully expressive and may require other functions in Jotai to support other use cases.

- Last but not least, the provider-less mode in Jotai, which allows omission of the `Provider` component, is technically simple, but very developer-friendly to lower the mental barrier as regards using the library.

Basic functionalities are the same in both Recoil and Jotai and developers would need to make a choice based on other requirements or just their preference in terms of the API. Jotai's API is minimalistic, the same as Zustand.

In this section, we saw a comparison between Recoil and Jotai. Next up, we will see a comparison between MobX and Valtio.

Using Valtio and MobX

Although the motivation is quite different, Valtio is often compared to MobX (`https://mobx.js.org`). Usage-wise, there are some similarities in Valtio and MobX regarding their React binding. Both are based on mutable states and developers can directly mutate state, which results in similar usage. JavaScript is based on mutable objects, so the syntax of mutating an object is very natural and compact. This is a big win for mutable states compared to immutable states.

On the other hand, there is a difference in how they optimize renders. For render optimization, while Valtio uses a hook, MobX React uses a **higher-order component** (**HoC**): `https://reactjs.org/docs/higher-order-components.html`.

In this section, we will convert a simple MobX example into Valtio. Then we will see the differences between the two.

> **Important Note**
>
> Conceptually, Valtio is comparable to Immer (`https://immerjs.github.io/immer/`). Both try to bridge immutable and mutable states. Valtio is based on mutable states and converts states to immutable ones, whereas Immer is based on immutable states and uses mutable states (drafts) temporarily.

Example involving MobX and Valtio

Let's take an example from MobX's documentation: `https://mobx.js.org/README.html#a-quick-example`.

We first import some functions from the MobX libraries:

```
import { makeAutoObservable } from "mobx";
import { observer } from "mobx-react";
```

As the MobX library is framework-agnostic, the React-related function is imported from the MobX React library.

The next step is to define the business logic, which is a timer. We create a class and then instantiate it:

```
class Timer {
  secondsPassed = 0;
  constructor() {
    makeAutoObservable(this);
  }
  increase() {
    this.secondsPassed += 1;
  }
  reset() {
    this.secondsPassed = 0;
  }
}

const myTimer = new Timer();
```

It has one property and two functions to mutate the property. `makeAutoObservable` is used to make the `myTimer` instance an observable object.

We can call the mutating function anywhere within the code. As an example, let's set an interval:

```
setInterval(() => {
  myTimer.increase();
}, 1000);
```

This will increase the `secondsPassed` property every second.

Now, the component to use `timer` is the following:

```
const TimerView = observer(({ timer }: { timer: Timer }) => (
  <button onClick={() => timer.reset()}>
    Seconds passed: {timer.secondsPassed}
  </button>
));
```

The `observer` function is an HoC. It understands `timer.secondsPassed` is used in a render function, and will trigger re-renders when `timer.secondsPassed` changes.

Finally, the `App` component has the `TimerView` component with the `myTimer` instance:

```
const App = () => (
  <>
    <TimerView timer={myTimer} />
  </>
);
```

As *Figure 11.3* shows, if you run this app, it will show a button with a label showing the number of seconds that have passed. The label changes every second. Clicking this button will reset the number.

Seconds passed: 7

Figure 11.3 – Screenshot of the app with MobX

Now, what would this look like with Valtio? Let's see the same example with Valtio.

We first import two functions from the Valtio library:

```
import { proxy, useSnapshot } from "valtio";
```

Although Valtio is a library for React, it has a vanilla bundle for non-React use cases.

We use the `proxy` function to define a `myTimer` instance:

```
const myTimer = proxy({
  secondsPassed: 0,
  increase: () => {
    myTimer.secondsPassed += 1;
  },
  reset: () => {
    myTimer.secondsPassed = 0;
  },
});
```

It has a `secondsPassed` property for a number value and two function properties to update the number value.

We use one of the function properties to increase the `secondsPassed` property periodically:

```
setInterval(() => {
  myTimer.increase();
}, 1000);
```

This `setInterval` usage is exactly the same as MobX.

Next is the `TimerView` component using `useSnapshot`:

```
const TimerView = ({ timer }: { timer: typeof myTimer }) => {
  const snap = useSnapshot(timer);
  return (
    <button onClick={() => timer.reset()}>
      Seconds passed: {snap.secondsPassed}
    </button>
  );
};
```

In Valtio, `useSnapshot` is a hook to understand how a state is used in a render function and will trigger re-renders when the used part in the state is changed.

Finally, the `App` component is the same as MobX:

```
const App = () => (
  <>
    <TimerView timer={myTimer} />
  </>
);
```

In the end, we should have the same behavior as MobX. It shows a button with a label. The label shows the number of seconds that have passed, and clicking the button will reset the value.

Now, let's discuss some differences.

Comparing examples of MobX and Valtio

The two examples in MobX and Valtio look similar, but there are two major differences:

- The first difference is the updating method. Although both use mutations, the MobX example is class-based, whereas the Valtio example is object-based. It's mostly stylistic, and Valtio is not very opinionated regarding the styles.

  One of the styles Valtio allows is the separation of functions from the state object. The same example can be implemented in the following approach:

  ```
  // timer.js
  const timer = proxy({ secondsPassed: 0 })

  export const increase = () => {
    timer.secondsPassed += 1;
  };

  export const reset = () => {
    timer.secondsPassed = 0;
  };

  export const useSecondsPasses = () =>
    useSnapshot(timer).secondsPassed;
  ```

  We define updating functions outside the state object defined by the `proxy` function. The benefit of this approach is that it allows code-splitting, minification, and dead code elimination. In the end, we can expect an optimized bundle size.

- The second difference is the render optimization method. While MobX takes the observer approach, Valtio takes the hook approach. There are pros and cons to each. The observer approach is more predictable. The hook approach is more "concurrent rendering" friendly. Implementing this approach is likely very different. There's also a stylistic difference; some developers prefer the HoC style, while other developers prefer the hook style.

> **Important Note**
>
> As of the time of writing, we only have limited information about concurrent rendering. It's our best observation at this point, but it's not guaranteed whether the statement will hold in the future.

In this section, we saw a comparison between MobX and Valtio. Next up, we will discuss a comparison between Zustand, Jotai, and Valtio.

Comparing Zustand, Jotai, and Valtio

In this chapter so far, we have compared the following pairs:

- Zustand and Redux in the *Differences between Zustand and Redux* section
- Jotai and Recoil in the *Understanding when to use Jotai and Recoil* section
- Valtio and MobX in the *Using Valtio and MobX* section

We compared these pairs because there are some similarities. In this section, we will compare Zustand, Jotai, and Valtio.

First of all, all three libraries are provided by the Poimandres GitHub organization (https://github.com/pmndrs). It's a developer collective providing many libraries. Three micro-state management libraries from a single GitHub organization may sound counter-intuitive, but they are in different styles. There is also a philosophy that is common in the three libraries: their small API surfaces. All three libraries try their best to provide small API surfaces and let developers compose the APIs as they want.

But then, what are the differences between the three libraries?

There are two aspects:

- **Where does the state reside?** In React, there are two approaches. One is the module state, and the other is the component state. A module state is a state that is created at the module level and doesn't belong to React. A component state is a state that is created in React component life cycles and controlled by React. Zustand and Valtio are designed for module states. On the other hand, Jotai is designed for component states. For example, consider Jotai atoms. The following is a definition of countAtom:

    ```
    const countAtom = atom(0);
    ```

    This countAtom variable holds a config object, and it doesn't hold a value. The atom values are stored in a Provider component. Hence, countAtom can be reused for multiple components. Implementing the same behavior is tricky with module states. With Zustand and Valtio, we would end up using React Context. On the other hand, accessing component states from outside React is technically not possible. We'll likely need some sort of module state to connect to the component states.

Whether we use module states or component states depends on the app requirements. Usually, using either module states or component states for global states fulfills the app requirements, but in some rare cases, using both types of states may make sense.

- **What is the state updating style?** There is a major difference between Zustand and Valtio. Zustand is based on the immutable state model, while Valtio is based on the mutable state model. The contract in the immutable state model is that objects cannot be changed once created. Suppose you have a state variable such as `state = { count: 0 }`. If you want to update the count in the immutable state model, you need to create a new object. Hence, incrementing the count by 1 should be `state = { count: state.count + 1 }`. In the mutable state mode, it could be `++state.count`. This is because JavaScript objects are mutable by nature. The benefit of the immutable model is that you can compare the object references to know whether anything has changed. It helps improve performance for large, nested objects. Because React is mostly based on the immutable model, Zustand with the same model has compatibility. Thus, Zustand is a very thin library. On the other hand, Valtio, with the mutable state model, requires filling the gap between the two models. In the end, Zustand and Valtio take different state updating styles. The mutable updating style is very handy, especially when an object is deeply nested. Revisit the example in *The pros and cons of this approach* section of *Chapter 9, Use Case Scenario 3 – Valtio*.

> **Note Regarding the Use of Immer**
>
> It's possible to use Immer to allow mutations to update states in Zustand and Jotai. Compared to the combination of Zustand and Immer, Valtio is more optimized for the mutable state model. It has smaller API surfaces and it also optimizes re-renders. The combination of Jotai and Immer is useful for big objects, and the Jotai library provides a specific feature to integrate Immer. However, Jotai atoms are usually small, and in such a case, the immutable updating style is not a big issue.

There are some minor differences among the three libraries, but what's important is the fact that they are based on different principles. If we were to choose one of them, we would need to see which principle fits well with our app requirements and our mental model.

Summary

In this chapter, we summarized the differences between the three libraries for a global state we explained in this book. They are different because they are based on different models.

Essentially, micro-state management involves choosing the right solution and the right library for a specific problem. Micro state management requires you to understand what your problem is and what solutions are available for your problem. We hope that this book has covered some topics that will help developers find the right solution.

Hi!

I am Daishi Kato, author of Micro State Management with React Hooks. I really hope you enjoyed reading this book and found it useful for increasing your productivity and efficiency in React Hooks.

It would really help me (and other potential readers!) if you could leave a review on Amazon sharing your thoughts on Micro State Management with React Hooks here.

Go to the link below or scan the QR code to leave your review:

https://packt.link/r/1801812373

Your review will help me to understand what's worked well in this book, and what could be improved upon for future editions, so it really is appreciated.

Best wishes,

Daishi Kato

Index

Z

Packt.com

Subscribe to our online digital library for full access to over 7,000 books and videos, as well as industry leading tools to help you plan your personal development and advance your career. For more information, please visit our website.

Why subscribe?

- Spend less time learning and more time coding with practical eBooks and Videos from over 4,000 industry professionals

- Improve your learning with Skill Plans built especially for you

- Get a free eBook or video every month

- Fully searchable for easy access to vital information

- Copy and paste, print, and bookmark content

Did you know that Packt offers eBook versions of every book published, with PDF and ePub files available? You can upgrade to the eBook version at packt.com and as a print book customer, you are entitled to a discount on the eBook copy. Get in touch with us at customercare@packtpub.com for more details.

At www.packt.com, you can also read a collection of free technical articles, sign up for a range of free newsletters, and receive exclusive discounts and offers on Packt books and eBooks.

Other Books You May Enjoy

If you enjoyed this book, you may be interested in these other books by Packt:

Simplify Testing with React Testing Library

Scottie Crump

ISBN: 978-1-80056-445-9

- Explore React Testing Library and its use cases
- Get to grips with the RTL ecosystem
- Apply jest-dom to enhance your tests using RTL
- Gain the confidence you need to create tests that don't break with changes using RTL
- Integrate Cucumber and Cypress into your test suite
- Use TDD to drive the process of writing tests
- Apply your existing React knowledge for using RTL

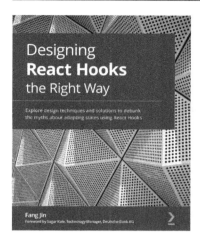

Designing React Hooks the Right Way

Fang Jin

ISBN: 978-1-80323-595-0

- Create your own hooks to suit your state management requirement
- Detect the current window size of your website using useEffect
- Debounce an action to improve user interface (UI) performance using useMemo
- Establish a global site configuration using useContext
- Avoid hard-to-find application memory leaks using useRef
- Design a simple and effective API data layer using custom Hooks

Packt is searching for authors like you

If you're interested in becoming an author for Packt, please visit authors. packtpub.com and apply today. We have worked with thousands of developers and tech professionals, just like you, to help them share their insight with the global tech community. You can make a general application, apply for a specific hot topic that we are recruiting an author for, or submit your own idea.

www.ingramcontent.com/pod-product-compliance
Lightning Source LLC
Chambersburg PA
CBHW060539060326
40690CB00017B/3549